Disasters, Culture, Politics

Disasters, Culture, Politics: Chinese-Bulgarian Anthropological Contribution to the Study of Critical Situations

Edited by

Elya Tzaneva (Editor-in-Chief),
with Fang Sumei and Liu Mingxin

CAMBRIDGE
SCHOLARS
PUBLISHING

Disasters, Culture, Politics:
Chinese-Bulgarian Anthropological Contribution to the Study of Critical Situations,
Edited by Elya Tzaneva (Editor-in-Chief), with Fang Sumei and Liu Mingxin

This book first published 2009

Cambridge Scholars Publishing

12 Back Chapman Street, Newcastle upon Tyne, NE6 2XX, UK

British Library Cataloguing in Publication Data
A catalogue record for this book is available from the British Library

ISBN (10): 1-4438-1348-6, ISBN (13): 978-1-4438-1348-8

TABLE OF CONTENTS

PREFACE

This collection was inspired by the Chinese-Bulgarian conference "*Contemporary Critical Situations: Natural and Social Disasters and Traditional Culture*" held in 2007, in Sofia. The conference was organized by the Ethnographic Institute and Museum at the BULGARIAN ACADEMY OF SCIENCES in Sofia (BAS), with the participation of two Chinese academic institutions: the Institute of Ethnology and Anthropology at the CHINESE ACADEMY OF SOCIAL SCIENCES in Beijing (CASS), and the School of Ethnology and Sociology – MINZU UNIVERSITY OF CHINA (MUN), former CENTRAL UNIVERSITY FOR NATIONALITIES, Beijing. The conference was realized as a Joint Research Project of the institutions mentioned within the framework of the General Agreement for Scholarly Cooperation between CASS and BAS, and came as a result of expanded joint field work and work-in-progress seminars in Beijing and Sofia, in the period between 2005 and 2007. In the conference took part, and contributed to this volume, thirteen recognized and respected ethnologists from the two countries, whose expertise and profound understanding of the problems arouse expectations of significant scholarly achievements.

As we have recently observed, it often happens that the typical pattern of relations between the individuals or groups and their natural and social environment is disturbed by unexpected circumstances or phenomena that are impossible to predict or even envisage. The catastrophic natural events – such as earth-quakes, floods, land-slides, tsunami-waves, fires – as well as those having technological and social background – such as epidemic diseases, consequences of wars, revolutions and migrations, oil-splits, construction disasters, explosions, etc. – always break the individual's and group's cycle of everyday culture and change its normal course. Recent perspectives in ethnological research define a disaster as a "process/event involving the combination of a potentially destructive agent(s) from the natural and/or technological environment and a population in a socially and technologically produced condition of vulnerability"[1]. The authors presented in the volume depart from the view that the cultural features of the population which is suffering a disasters or a crisis forms a specific

[1] A.O.Smith. Anthropological Research on Hazards and Disasters. *Annual Review of Anthropology*, 1996, No25, p.303-328.

pre-disaster vulnerability of the society. The interrelationships between cultural, demographic, political, economic, and environmental spheres – that is, the social context of the crisis – define the preparedness, mobilization, and prevention of disasters for each discrete group of people or society. The understanding of local communities' social experience is, therefore, a significant part of the overall strategy for going through, surviving and recovering from the crisis.

Based on this general understanding, the present volume investigates a variety of situations of disaster dealt with in two contemporary societies in the increasingly globalizing world – that of China in the East Asia, and of Bulgaria in the Southeast European Balkans. Over the years, the modern Chinese have experienced many a natural anomaly or social cataclysm; especially dramatic was their facing and fighting SARS in 2003, and the Sichuan Earthquake in 2007. In the long run, of great importance for China is the overcoming of disasters with longer life-cycles, crises "in progress", such as climatic and bio-disasters, complex problems associated with the arid areas, spread of epidemic diseases etc. Over the last decades Bulgarian society has been hit by different type crises caused by the new political development in the country – the post-socialist period and the transition to European democratic institutions; but there have also occurred draughts, extreme snow-falls; in 2006 there was an avian influenza threat which caused great anxiety and asked for maximal mobilization of physical and cultural resources of society. The authors lay special stress on the role of the traditional cultural systems (the beliefs, behaviors, and institutions characteristic of a particular society or group) in a situation of disaster; a number of articles focus on how people draw upon and alter their belief systems over longer periods of time in their efforts to come to terms with catastrophic changes, violence, loss, resettlement, and even humanitarian relief. Studied is the role of such events in the changes occurring in social institutions such as religious beliefs or customs, social organizations, attitudes and values.

The scholars from the two ethnological traditions presented in the volume have contributed to the understanding of these phenomena. In the materials of both Bulgarian and Chinese ethnologists, some of the adaptive coping strategies are shown traditionally used even by relatively distant world populations to respond and cope with crises. The ethnologists have worked on the assumption that through studying the collective responses to disasters in the two respective cultures they could also contribute to the understanding of their causes. Concluding that disasters and disaster relief can dramatically impact the material conditions of life of the affected population, some authors have studied the changes that disasters can bring to economic systems and related mechanisms. Also studied are the ways in which the critical situations can alter political organizations and power relations between individuals, the state, and the international institutions. The articles reveal that, having gone through and survived a disaster, people more clearly comprehend their own political place and situation and their own position of power (or lack of such) relative to that of the state and all official representatives and authorities involved.

The volume applies a broad ethnological approach in the study of disasters, the matter being interpreted comparatively, contextually, and in cross-cultural perspectives, which is all the more justified in view of the fact that it represents an extensive work in two non-western, therefore, lesser known in the discussed perspective cultures of the modern world.

The most significant contribution, however, is believed to be the compilation and analysis of unique empirical data and their evaluation according to the theoretical assumptions of the members of the joint research team. The expected results and benefits from the study are in the fields of both scientific prognostics and social practice. It could give an answer to some important questions, such as: In what way and to what extent do the extreme natural and social situations affect the lifestyle, spirituality and ideology of the population concerned? How do the stereotypical thoughts and acts of the population change, when the familiar environment is destroyed or damaged by a disaster? What is the place of traditional values in critical circumstances? How do societies create their own vulnerabilities? Is it possible to develop a model for successful behavior for coping with disasters based upon cultural traditions? What is needed for most successful relief management?, etc. On a broader methodological level, the analysis aims at finding a balance between economic, political and cultural factors for prevention and overcoming of disasters.

We believe that the focus on such important social matters will even lead the two academic teams to achieving results and reaching conclusions that will cross the two-culture scheme of investigation and will contribute to expanding the geographical and thematic scope of the study. The two partner Academies – CASS in Beijing and BAS in Sofia – have reached an agreement for further investigations within the subject of disasters and crises, and the joint Chinese-Bulgarian scholarly team is now organizing the second stage of the research, challenged by the recent catastrophic events intensified in number and going deeper in significance. This first issue is not meant to give a definitive statement on the topic. The authors hope that in fact, if it is successful, the volume will identify more scientific problems than it has solved.

As an initiator and organizer of the first Chinese-Bulgarian anthropological joint research team and an editor-in-chief of the volume I would like to express a special gratitude to two Bulgarian scholars, professors in History, who have stimulated the efforts of the researchers with their constant help and encouragement, namely, Prof. Konstantin Kossev, Vice-President of the Bulgarian Academy of Sciences, who has signed the General Agreement between the two Academies in 2001 within which the project was carried out, and Prof. Dimitar Tzanev, Ambassador of Bulgaria in China (1999-2003), whose motivation and personal involvement in the idea of Chinese-Bulgarian academic contacts gave the original and main impetus to the realization of this project. We are also thankful to the International Departments of the institutions involved – Bulgarian Academy of Sciences, Chinese Academy of Social Science, and Central University for Nationalities (now Minzu University of China) for their professional assistance.

—Elya Tzaneva

QINGHAI PROVINCE IN THE PERSPECTIVE OF ECOLOGICAL ANTHROPOLOGY: ECOLOGICAL CRISIS, ECONOMIC DEVELOPMENT AND THE PRESERVATION OF CULTURAL DIVERSITY

HAO SHIYUAN

PROFESSOR, DIRECTOR OF THE INSTITUTE OF ETHNOLOGY AND ANTHROPOLOGY, CHINESE ACADEMY OF SOCIAL SCIENCES

Since the 1990s, a series of global issues have caused widespread concerns around the world. Among these, of special interest is the issue of ecological environment, which typically involves the conservation of biological diversity. In the UN Summit Conference on Environment and Development held in Rio de Janeiro in 1992, 156 countries and the European Community signed the *Convention on Biological Diversity*, so as to protect the habitats, species and Community genetic resources, to achieve the sustainable use of resources, and to ensure a fair and equitable sharing of benefits arising from biodiversity via policy, economic and managerial channels. In that November, China became one of the first six countries in the world that ratified the Convention.

The term "biological diversity" or "biodiversity" was initiated in the early 1980s. According to the "Convention on Biological Diversity", biodiversity refers to: "the variability among living organisms from all sources including, terrestrial, marine and other aquatic ecosystems and the ecological complexes they compose; this includes diversity within species, between species and of ecosystems."[1] With its diverse ecosystems, China ranks 3rd in the world with its exceedingly rich floral resources, over half of which grow nowhere else. China also boasts of one eighth of the world's mammals, birds and fish, and is regarded as one of the centers for

[1] State Environmental Protection Administration, China's Biodiversity: Country Study, Beijing, 1997, p.11.

genetic variants of domesticated crops in the world[2]. The diverse biological resources not only endow China with the advantage of achieving a coordinated development between man and nature, but will produce great impacts on global environment and human society at large.

Early back in the 1950s, some Chinese scholars proposed the initiative of establishing nature reserves. The first nature reserve (Dinghushan) was established in Guangdong Province in 1956.[3] In the following half century, China's nature reserves witnessed steady development. Since the reform and opening up particularly, people's awareness of the environment and ecological conservation has been enhanced; in October 1994, the *Regulation on Nature Reserves of China* was adopted by China, which was followed by the *Development Plan of Nature Reserves in Qinghai Province (1996-2010)* in 1996. The Plan provides that "from 2001 to 2010, the total number of nature reserves should reach 1200 (including 160-170 national nature reserves), and the total area of nature reserves should account for 10% of the overall territory of China".[4] As of August 2005, China had established over 2000 nature reserves (of which 243 were national), making up 15% of the total territory. Obviously, China has made important progress in nature reserve development and biodiversity conservation.

However, as found elsewhere in the world, the enhancement of biodiversity conservation consensus is always accompanied by the accelerating extinction of species and the ensuing ecological imbalance. In China, due partly to its large population and partly to the growing demand of economic and social development for natural resources, there exists a serious threat to the sustainable utilization of biodiversity resources. Still, the climate impacts of global warming further aggravate the ecological environment issue. China is gradually losing its advantage of rich biodiversity. Such examples include: the sharp reduction and fragmentation of forest areas, overgrazing and severe degradation of grassland, predatory animal hunting and herb gathering, over-fishing of hydrobios, mining activities, environmental pollution and the invasion of alien species, which have all posed a threat to China's biodiversity resources, such that the list of extinct or endangered species in China grows longer and longer.

---

[2] Council for International Cooperation on Environment, International Environmental Cooperation and Sustainable Development: Conserving China's Biodiversity, Beijing, 1997, p.3.

[3] General Report Compile Group of China Biodiversity Conservation Action Plan, China Biodiversity Conservation Action Plan, Beijing, 1994, p.16.

[4] Outline of Plan for Nature Reserves Development in China (1996-2010), approved by the State Council, State Environmental Protection Administration, State Planning Committee, issued on November 24, 1997.

This is particularly so with Qinghai Province in the Tibetan plateau. As revealed in the *Environmental Condition of Qinghai Province in 2004*, in most areas of the province, tendencies of vegetation degradation, land desertification, soil erosion and the destruction of wetlands and rare wildlife habitats have not been brought under control. In June 2005, the Qinghai Environmental Protection Bureau also noted that vegetation degradation, wetland shrinkage and expansion of desertified land have led to a deterioration of the living environment of wildlife and a reduction of their distribution area, and that the excessive animal hunting and herb gathering (such as Chinese caterpillar fungus) have caused the plants and animals in the Tibetan plateau to decline sharply, and have brought some species to the brink of extinction. The endangered biological species account for 15%-20% of the total in the province, obviously higher than the world average level of 10%-15%.[5] For this reason, Qinghai Province has become one of the hot-spots for ecological conservation research both at home and abroad.

I. Nature Reserves in Qinghai: An Introduction

With its vast territory, various landforms and complex ecological environment, the Tibetan plateau has a uniquely rich collection of species. Statistics indicate that there are over 13 000 known species of higher plants (including mosses) in the Tibetan plateau, and nearly 1100 kinds of terrestrial vertebrates, each accounting for about 45% of the total number in the whole country. In addition, there are 115 kinds of fish, 5000 kinds of fungi, and a huge number of insects, invertebrates, algae, lichens, bacteria, viruses, and so on. In accordance with the nine major classifications of nature reserves of China, the area is designated as "The Cold Desert Area in the Tibetan plateau," characterized by sparse population and slow development. That is, it is suitable for the establishment of nature reserves.

As part of the Tibetan plateau, Qinghai Province covers a total area of 722 300 square kilometers. Qinghai Province has complex and varied landforms, with an average altitude of 3000 meters above sea level, ranging from the highest 6860 meters to the lowest 1600 meters. Qinghai Province also has diverse geographical landscapes, such as glaciers, mountains, gobies, deserts, rivers, lakes, basins, grasslands, forests, salt lakes, hungriness, and so on (Illustration 1). The average annual rainfall is below 400 mm; the weather here is typically cold and dry, and the annual

[5] See: "Endangered Species in Qinghai Province is Higher than the World Level", Qinghai News Net, June 27, 2005, http://www.sina.com.cn.

average temperature is -5.7°C–8.5°C. The lands in Qinghai Province are under-developed, with only 6814.34 square kilometers (0.94%) of arable land and 24 240 square kilometers (3.4%) of forest land, as compared with the 406 643 square kilometers (56.35%) of grassland area. The water area of the province is 31 615 square kilometers (4.38%), and the barren land area is 250 021 square kilometers (34.65%).

Qinghai Province has a wide assemblage of biological resources, with about 3000 kinds of plants, 103 kinds of beasts, 294 kinds of birds, 56 kinds of fish, and so on. Of these, 110 species of plants and animals are on the list of national protection. For all that, the ecological environment in Qinghai is vulnerable. The forest coverage rate of the province is far below the national average level of 13%. Besides, these forests are unevenly distributed across the limited forest growth zone, and are thus irreversible once damaged. While at the same time, due to the limited arable land in this area, agricultural production is hard here. Although the grassland coverage rate is relatively high, it is basically dry and cold. Still, the long withering period in winters and springs imposes a constraint on the development of animal husbandry. In addition, the frequent natural disasters in this area such as droughts, snowstorms, hails, frosts and earthquakes can bring great harm to the natural environment. This situation was further exacerbated by the ever expansion of human activities.[6] Therefore, economic and social development in such an area is completely different from that in the central and eastern regions, as safeguarding the security of ecological environment is the basic premise for the sustained economic and social development of the Tibetan plateau.

The development of nature reserves is of great significance to the maintenance of the ecosystem and biodiversity in the region. By the end of 2006, 8 nature reserves (of which 5 are national-level) had been established in Qinghai Province, with a total coverage of 206 100 square kilometers, accounting for 28.53% of the Qinghai territory, which is higher than the national average level.

[6] T. J. Wang. Population and Environment Carrying Capacity on Qinghai-Tibet Plateau, Beijing, 1998, p.25.

Table 1: National Nature Reserves in Qinghai Province[7]

Names of Reserves	Administrative Regions	Areas (10 000 hectares)	Main Protection Targets	Time of Approval
Qinghai Lake National Nature Reserve	Hainan Prefecture and Haibei Prefecture	49.52	Bar-headed Goose,brown -headed gulls and other water birds and ecosystems	Dec., 1997
Kokoxili National Nature Reserve	Yushu Prefecture	450.0	Tibetan antelope, wild yak, wild ass, and snow leopard and other wild animals	Dec., 1997
Yushulong National Nature Reserve	Yushu Prefecture	1.0	Black-necked crane, swan and other waterfowl and plateau wetland ecosystem	Aug.,1986
Mengda National Nature Reserve	Haidong regions	1.729	Forest ecosystems and rare species	Apr., 2000
Sanjiangyuan National Nature Reserve	Yushu Prefecture, Guoluo Prefecture, Hainan Prefecture, Haixi Prefecture, and Huangnan Prefecture	1523.0	Forests, wetland ecosystems and wild animals	Jan., 2003

7 See: http://www.qhepb.gov.cn/qhepb/index.asp.

The nature reserves in Qinghai Province have 40 years of history. In addition to the five state-level nature reserves, there are 46 nature reserves of provincial, municipal or national levels, 12 national and provincial forest parks and 19 game sanctuaries. According to the *Development Plan of Nature Reserves in Qinghai Province (1999-2010)*, 10 more national nature reserves and 8 provincial nature reserves will be built by 2010, with a newly-added area of 8.229 hectares. Except a few upgraded ones, all of the nature reserves are newly established or reestablished. However, due to the global warming and intensification of human economic activities, the ecological environment problem with these nature reserves stands out as peculiarly glaring.

II. Ecological Environment Problems with Qinghai Nature Reserves: Case Study

In 2005, I conducted an investigation on the ecological environment crisis in Qinghai, focusing on the Hol Xil National Nature Reserve and the Qinghai Lake National Nature Reserve.[8] I found a number of serious problems with the ecological environment in these areas as a result of climate change and human activities:

(I) Hol Xil National Nature Reserve covers an area of 45 000 square kilometers, about half of which is located in the core area. The largest and highest of its kind in China, the reserve enjoys a most diverse collection of wildlife resources. For this reason, it is widely known as "the third pole of the world" and the gene pool of rare wild animals on the Tibetan plateau. Among others, Tibetan antelope is one of the most precious wild animals (Illustration 2). From the 1990s onward, continued illegal poaching of Tibetan antelopes has been going on in Hol Xil, so as to obtain antelope skins and cashmere and smuggle them abroad for huge profits. Accordingly, there has emerged an international crime network with Kashmir as its center. From Kashmir, antelope products are sold out to various fashion centers all over the world. In response to the rare animal crisis, the Chinese Government has taken strong measures against it, either by establishing law enforcement agencies or improving management methods, which have effectively restrained the illegal poaching. And during the construction of Qinghai-Tibet railway through the Hol Xil region, a lot of elevated railway bridges are built so as to ensure the traditional migration routes of

---

[8] S. Y. Hao. Qinghai-Tibet Plateau Biodiversity Conservation and Environment – Taking the Qinghai Nature Reserve as an Example, Investigation Report, 2005.

Tibetan antelopes and other wildlife. These measures have provided an important guarantee for the safety of the Hol Xil Nature Reserve, and the number of Tibetan antelopes and other wild animals is seeing rapid recovery.

However, due to the macro effects of global warming, the rodent problem is more keenly felt than ever in the cold and high meadow areas of Hol Xil Nature Reserve. Rat holes keep growing, and large expanses of turf maintaining the vulnerable eco-system of the reserve are liable to rodent damage (by plateau voles, mice, rabbit and other rodents) and wither away, which has led to the degradation, even desertification, in some regions of the grassland. Since plateau voles have high reproductively in high and cold regions, with an annual assumption of about 50 kg forage for each vole, hence the food-vying problem with the growing number of Tibetan antelopes, Tibetan gazelle, wild yak, Tibetan wild ass and other wild animals. Meanwhile, as some of the nearby herdsmen enter the reserve to graze their cattle in better pastures, tens of thousands of livestock occupy some of the habitats and water supplies of Tibetan antelopes, affecting the original ecological environment of Tibetan antelopes and other wild animals. Meanwhile, the co-existence of domestic and wild animals also opens up the possibility for their mutual mating, which may, to some extent, cause the variation of highland wildlife species and endanger the originality of the species, genes and genetic aspects of biological diversity.[9]

Therefore, although steady progress has been made in infrastructure, protection awareness, protection measures and protection effectiveness, the Hol Xil Nature Reserve is under the dual challenge of "natural disasters" and "man-made disasters". Since October 1, 2000, fierce measures have been taken against grazing, hunting, herb gathering, mining, gold mining and other activities in the Hol Xil Nature Reserve, nevertheless, there is still a long way to go before these measures can really take effect.

(II) As China's largest salt lake, the Qinghai Lake National Nature Reserve covers a water area of 4635 square kilometers and has a water volume of 105 billion cubic meters. It is specially oriented for waterfowl and wetland ecosystems. There are 14 orders, 33 kinds and 189 species of birds in the nature reserve. It is the first nature reserve established to protect the waterfowl and wetland ecosystems in Qinghai Province, and is listed as one of the Important Wetlands in the World.

---

[9] See: "Hol Xil Depopulated Area is facing the Dangerous Ecological Threat", June 28, 2005, Xinhuanet.com.

The major source for Qinghai Lake comes from dozens of its inflow rivers, along with the infiltration of groundwater from surrounding mountains. Qinghai Lake is both an important water body to maintain the ecological security of the northeastern Tibetan plateau and a natural barrier to block off the eastward desertification. It is also one of the natural bases for the habitat and reproduction of 8 major birds in China. Since the establishment of the Qinghai Lake Nature Reserve, the ecological environment of the lake region and bird biodiversity have been put under protection. Over the years, however, due to the impacts of global warming and excessive human activities, the continued deteriorating tendency has not eased. In terms of the effects of climate change, the total water supply to the Qinghai Lake each year is 3.493 billion cubic meters, including 1.335 billion cubic meters of runoff supply, 1.557 billion cubic meters of precipitation supply and 401 million cubic meters of groundwater supply. And yet, given that the annual evaporation of the lake is 3.930 billion cubic meters, the Qinghai Lake actually suffers an annual water loss of 437 million cubic meters. The main reason for the water-level drop of Qinghai Lake is drought and severe evaporation, as a result of the changes in ecological environment. Monitoring data indicate that the water level of Qinghai Lake dropped by 11.7 meters in 1908-2000[10]. In the past half century, the water area of Qinghai Lake has dwindled by 362.3 square kilometers[11]. The changes in climate and environment include the drying-up of the inflow water sources in the lake region, and the continued dropping of water level has caused the salt and alkali ingredients to increase sharply[12].

Another sign for the changes in the water body and environment of Qinghai Lake is the continued expansion of desertified region around the lake, largely as a result of the illegal reclamation of grassland and rape-growing in the 1990s. At the same time, over-grazing around the lake also accelerates the degradation of grassland around the lake. As the lake-based agricultural development draws heavily on the water from the inflow rivers for irrigation, Qinghai Lake suffers further losses. In short, the changes caused by natural factors and human activities have seriously affected the biodiversity of the lake. Statistics indicate that the species

[10] See: "Qinghai Lake: Water Level Dropped Mainly Due to Natural Disasters", http://www.cws.net.cn /cwsnews/newshtm/qhh/.

[11] See: "Ecological Environment Protection and Comprehensive Management Project of the Qinghai Lake Watershed Has been Started", http://news.sohu.com/20080526/n257087046.shtml.

[12] See: "An Important Sign for the Deterioration of the Ecological Environment: the Qinghai Lake becomes Salter", http://www.cws.net.cn/cwsnews/newshtm/qhh/.

under threat in this region account for about 15-20% of the total. Gymnocypris Orzewalskii is a typical example here.

Qinghai Lake is the main producing area of Gymnocypris Orzewalskii. Buha River is the largest inflow river of Qinghai Lake, accounting for 70% of the annual inflow water of Qinghai Lake. Moreover, it is also the most important spawning base for Gymnocypris Orzewalskii. However, the water volume of Buha River is reducing year by year due to a reduction of rainfall. Each year, a large number of countercurrent spawning Gymnocypris Orzewalskiis die of grounding and are incapable of return ing to lake. At the same time, due to the low water temperature, high salinity, high alkalinity and lack of natural baits, Gymnocypris Orzewalskii are seriously affected in terms of feeding, growing and reproducing. In addition to the abominable natural disasters, excessive human activities such as over-fishing can also bring about disastrous consequences to Gymnocypris Orzewalskii. Continued (over-) fishing in spawning areas has resulted in the fact that the number of Gymnocypris Orzewalskii in Qinghai Lake is only one tenth of that during the early period in the 1960s.[13] In 1986-2001, the Qinghai provincial government promulgated four fishing bans, which, in a sense, reflect the severity of the over-fishing problem and the difficulty of lake enclosing and fish cultivation. The lake enclosing and fish cultivation period adopted in 2001 is 10 years.[14] Zoologists believe that Gymnocypris Orzewalskii is a very important link in the ecological chain of Qinghai Lake. The exhaustion of Gymnocypris Orzewalskii resources will not only damage the ecological balance of "coexistence of fish and birds" in Qinghai Lake, but will threaten the survival of birds (It is estimated that about 1000 tons of Gymnocypris Orzewalskii are eaten by birds every year), which will in turn affect the overall ecology of the entire Qinghai Lake. In fact, negative consequences are looming; the perching and multiplying of birds in recent years has been directly influenced by the environmental degradation, and the unknown diseases are also quite serious.

[13] See: "Qinghai Lake: the Anxious Gymnocypris Orzewalskii and Three Major Dangers", http://www.cws.net.cn/cwsnews/newshtm/qhh/.

[14] See: "Save the Qinghai Lake", http://www.cws.net.cn/cwsnews/newshtm/qhh/25.html.

III. Economic Development and Ecological Environment Conservation

China is a developing country with the fastest economic growth in the world. The rapid economic and social development bears directly on its ecological environment and natural resources. Since Qinghai Province is lagging behind in economic and social development, its ecological environment is vulnerable too. According to the statistics from the Chinese Academy of Sciences, the 2007 ranking of Qinghai Province in the sustainable development index system in China is as follows:

No27 in terms of the level of regional social development, No28 in terms of the level of regional social security, No28 in terms of the driving force for regional social progress, No23 in terms of regional education capacity, No29 in terms of regional science and technology capacity, and No29 in terms of regional management capacity. As shown in the ranking, Qinghai Province is a typical example of economic and social underdevelopment. In terms of ecological and environmental conditions, Qinghai Province ranks No28 in regional ecological level, which consists of 3 components: ecological vulnerability, climate variability, and soil erosion. Its general ecological situation is very poor. The only index that Qinghai Province takes the lead (No3) is the regional environmental level that consists of industrial discharge and air pollution.[15] However, the index all the more indicates the industrial underdevelopment of Qinghai Province. Industrialization is the only way to modernization, but in the contemporary world, particularly in those developing countries where modernization is picking up speed, industrial discharge and the subsequent air pollution have become the major problem have aroused global concerns. China is no exception.

In the 21st century, with the acceleration of western economic and social development in China, the industrialization of Qinghai has witnessed significant expansion. However, in order to achieve industrial development in ecologically vulnerable areas such as Qinghai, ecological conservation and environment administration must be strengthened. Table 2 shows the ups and downs of the generation and discharge of industrial solid wastes in Qinghai Province in recent years.

---

[15] Sustainable Development Research Group of China, CAS, China Sustainable Development Strategy Research Report 2007, Beijing, 2007, p.410-425.

Table 2: Industrial Solid Wastes Generated & Discharged[16]

Year	Industrial Solid Wastes Generated (10 000 tons)	Industrial Solid Wastes Discharged (ton)
2003	379	75619
2004	508	60516
2005	649	27625
2006	882	3154

As shown above, the amount of industrial solid wastes in Qinghai Province in 2003-2006 increased year by year, indicating the scale and speed of economic development. At the same time, the year-on-year (and even drastic) decrease in the discharge of industrial solid wastes reflects a strengthened administration of the environment. In addition, Qinghai also set about restoring the ecological environment in nature reserves under the state support. For example, China launched the Sanjiangyuan ecological conservation project in 2005, with an initial investment of 7.5 billion Yuan (about 100 million USD) for ecological conservation and restoration in Qinghai.[17] And the administration work in the Hol Xil Nature Reserve also began to have effects on the problem of poaching, such that the scale of Tibetan antelopes is in gradual recovery, and the number of Tibetan antelopes has grown from 70 000 to 150 000 in the nature reserves of Qinghai, Tibet and Xinjiang. Meanwhile, remarkable progress has also been made in the administration of the Qinghai Lake Nature Reserve.

On August 1, 2003, the *Conservation Regulation of Ecologic Environment in the Qinghai Lake Watershed* (Draft) was adopted by Qinghai Province as the first regulation for nature reserves in China. This indicates that the ecological environment of Qinghai Lake and its watershed has come under legal protection. The regulation identifies that the goal of the ecological environment conservation of Qinghai Lake watershed is to maintain the biodiversity and preserve the natural eco-system of Qinghai Province. It focuses on the preservation of water bodies, wetlands, vegeta-

---

[16] Source: www.chinagate.com.cn.

[17] "Sanjiangyuan" means - 49.2% of the total flow of Yellow River, 25% of the total flow of Yangtze River and 15% of the total flow of Lancang River are from this region, which is known as "Water Tower of China". As part of the Qinghai-Tibet Plateau, the unique geographical environment and climatic conditions of "Sanjiangyuan" has extremely important impacts on the atmospheric circulation not only in China but also in East Asia and even the northern hemisphere.

tion and wild animals by addressing the relationship between the ecological environmental conservation on one part and economic development and the interests of farmers and herdsmen on the other part. In May 2008, the Ecological Environment Conservation and Comprehensive Administration Project of the Qinghai Lake watershed was initiated, with a total investment of 1.567 billion Yuan and a trial implementation period of 10 years. The project is mainly concerned with such issues as wetland conservation, administration of degraded grassland, prevention against rodent and insect pests, desertified land control, and returning farmland to forage land, ecological migration, etc. (Illustration 3). It also includes the reconstruction of six inflow rivers (200 km) such as Buha River and the removal of a power plant dam.[18]

One of the most important measures for the administration and conservation of the ecological environment of nature reserves is to minimize man's economic activities. Therefore, during the implementation process of the ecological restoration and environmental administration of Sanjiangyuan, Hol Xil and Qinghai Lake nature reserves, the relocation, or ecological migration of farmers and herdsmen in the nature reserves has emerged as a new way of migration.

IV. Ecological Migration and the Protection of "Two Kinds of Resources"

The term "ecological migration" has dual senses. The first sense refers to the migration activities caused by the deterioration of natural environment that has become unfit for human existence. The second sense has to do with the migrant measures taken to protect the ecological environment and reduce human activities. While the former often occurs naturally in the history of mankind, the latter is mainly carried out under state guidance. In contemporary China, ecological migration has become an important policy of achieving sustainable development. In the western areas inhabited by minority nationalities in particular, ecological migration has become an important means of preserving the ecological environment, improving people's livelihood, and promoting urbanization. The ecological migration in Qinghai Province is a typical one. The purpose of ecological migration is to ensure that no human activities occur within the bounds of nature

[18] See: "Ecological Environment Protection and Comprehensive Management Project of the Qinghai Lake Watershed Has been Started", http://news.sohu.com/20080526/m257087046.shtml.

reserves, so that the damaged ecological environment can be restored to its original functions.

Ecological migration constitutes an important part of the ecological construction plann of the three national nature reserves in Qinghai Province. In the core zone of Hol Xil Nature Reserve, the basic objective of ecological environment and biodiversity conservation is to achieve a "man-free zone". In the Sanjiangyuan Nature Reserve, ecological migration started in 2003. And by the end of 2007, a total of 13 305 households and over 60 000 people had been relocated or resettled.[19] In the Qinghai Lake Nature Reserve, the 2008 ecological environment conservation and administration project also includes a migration plan of 881 households and 4157 people, which can reduce nearly 1 million sheep units of pasturage.[20] Such ecological migrations do not merely mean a change in traditional residence, but also a change in production and lifestyle; urbanization has become the basic model of relocating ecological migrants.

At present, in terms of the ecological migration in Qinghai Province, great improvement has been made in the living conditions of residents, children's education, health care, transportation and communications. Moreover, the resettlement of ecological migrants can entail a shift from rural and pastoral areas to cities and towns, i.e. their production and lifestyle have changed dramatically during the migrations. Their economic pattern also undergoes a change from traditional animal husbandry into a variety of business services, such as small-scale processing. The migrants are mainly engaged in textile and blanket weaving, herb gathering, transportation, trade, labor export, and so on. These changes bear directly on their production and lifestyle. Moreover, there is the issue of transition and adaptation of traditional cultures. This is exactly the common concern of Chinese ethnographers and anthropologists.

With its cultural diversity, Qinghai is an important province with multi-ethnic population in western China. In 2005, the total population of Qinghai Province was 5.42 million, of which there were 1.191 million Tibetans, 863 000 Huis, 225 000 Tus, 119 000 Salars and 95 000 Mongolians, accounting for 46.32% of the total. Most of these ethnic minorities are engaged in animal husbandry, agriculture and commerce; they believe in Tibetan Buddhism and Islam. What is unique with these ethnic cultures is that they all rely on the ecological environment. As the Chinese folk

---

[19] See: "Qinghai: China's Largest Ecological Migration Project Progressing Smoothly", http://www.ce.cn/xwzx/gnsz/gdxw/200710/05/ .

[20] See: "Ecological Environment Protection and Comprehensive Management Project of the Qinghai Lake Watershed Has been Started", http://news.sohu.com/20080526/n257087046.shtml.

saying goes, "Those living on a mountain live off the mountain, those living near the water live off the water." After all, human existence and development would not have been possible without natural environment. It is the biological instinct of man to exploit various natural resources for his livelihood and reproduction. And it is the biological intelligence of man to transform nature for inventions and creations. However, his transformations and inventions are not isolated from the geographical environment and ecological resources, in the same way as boats are always invented near rivers and lakes, and skiing cannot be invented in Africa. This is one of the major problems that have drawn the attention of ecological anthropology, since "the theme of environmental determinism has basically been replaced by the people-environment model."[21]

Nowadays, modernization has become the common goal pursued by all mankind and by all nations, and human society faces the cultural and ecological problems, calling out for a new concept of development characterized by the interaction between man and nature. This concept of development is a preservation of both biodiversity resources and cultural diversity resources. That is to say, man faces the dual mission of preserving "two kinds of resources" in terms of development. An important lesson from western modernization is that "the development concept looks at things from the aspects of economics and quantity, and always ignores the cultural heritage of the ancient or traditional society."[22] As a result, many traditional wisdoms are lost before they are fully understood, scrutinized and assimilated. Cultural diversity is a valuable resource for human society, "from the historic point of view, human cultures tend to well adapt to and effectively promote the stability and vitality of its surrounding environment."[23] As far as China's minority regions are concerned, a lot of production experiences adapted to local conditions and knowledge maintaining the ecological balance manifest themselves in mountain agriculture, irrigated agriculture, grassland animal husbandry and hunting industry in the forests. These traditional wisdoms and values are one of the important intellectual sources for the enrichment and implementation of the new concept of development, and the discovering, integrating and developing of this knowledge is not only the basic prerequisite for the conservation and utilization of the "two kinds of resources", but also a necessary condi-

[21] D. L. Hardesty. Ecological Anthropology, translated by F. Guo, H. Zou, etc., 2002, p.3.
[22] E. Morin & A. B. Kern. Terre-Patrie, translated by S. L. Ma, 1997, p.80.
[23] D. A. Coleman. Eco-Politics, Building a Green Society, translated by J. J. Mei, 2002, p.117.

tion for respecting knowledge, motivating social vitality and achieving creativity.

Based on such an understanding, people began to realize that "the development should be designed in the way of anthropology, and the real development is the development of human."[24] As a subject that studies the human living and development, anthropology "has begun to affect the area of practical affairs. Governments and international institutions begin to find that to accept anthropologists' public policy, which is familiar with the social life, is beneficial."[25] But, of course, the anthropology here does not refer to the anthropology in the traditional sense, but the anthropology with the aim of human equality, coexistence and development, and of the conservation, heritage and development of cultural diversity. The civilized view of development based on the relationship between man and nature should "include the existing experiences of many civilizations, the complex network of natural world, the thorough understanding of the human history, and scientific research achievements." It also includes the absorption of the scientific theories and methods of ecological environment. Such a view should receive more attention in western regions such as Qinghai Province, namely, the traditional wisdom of the benign interaction between man and nature in traditional minority cultures.

The purpose of ecological migration is to recover and restore ecological environment. Meanwhile, ecological migrations, particularly those involving fundamental changes in production and lifestyle, may result in the loss of many ancient wisdoms and unique cultures. The changes from nomad to settlement, and from husbandry to trade and commerce, have actually invalidated some of the traditional knowledge of migrant groups that has been going down for generations. As a result, they are easily ignored and forgotten. However, such knowledge is a valuable cultural heritage for a nation, a society as well as the whole human civilization, in that the knowledge contains in itself man's experiences about his harmonious coexistence with nature. For example, the nomadic mode of production is rich in the traditional knowledge about preserving grassland and maintaining species balance. Therefore, how to draw on the ancient wisdom in traditional cultures is the important part of implementing the new view of development.

While at the same time, for the migrant groups who have entered trade and commerce, processing industry and handicraft industry in urban areas, a series of characteristics embodying cultural diversity such as language,

---

24 E. Morin & A. B. Kern. Op.cit., p.109.

25 M. Carrithers. Why Humans Have Cultures - Explaining Anthropology and Social Diversity, translated by F. Chen, 1998 version, p.206.

customs, clothing, diets, living artifacts, religious beliefs and rituals also need to be protected and developed. This is especially so when these migrant groups are integrated into an urban social environment of cultural diversity, and their cultural adaptation will give rise to a series of problems of social relations, or even the contradiction and conflicts between ethnic groups. Therefore, recognizing cultural differences and keeping them in harmonious coexistence for common development is an important problem confronting human society[26].

In the course of economic and social development and building a moderately prosperous society in China, the scientific outlook on development put forward by the Chinese government is to seek a sustained mode of development characterized by the benign interaction between man and nature. In order to implement the scientific outlook on development, we need to reconsider and recreate a series of issues, such as the development concept, development mode, development experience, etc. Meanwhile, the Chinese Government set the goal of building a harmonious society and proposed the important concept of "respecting differences and recognizing varieties". "Differences" and "varieties" are the basis of contradictions, but they do not necessarily lead to contradictions. The key issue is how to deal with the "differences" and "varieties". "Respecting differences" is the prerequisite for achieving a harmonious coexistence, and "recognizing varieties" is the guarantee of realizing complementary innovation. Countries with different civilizations and cultures should "make up for each other's deficiencies in their competitions and comparisons, and achieve common development by seeking common ground while reserving differences", which is the central tenet of "respecting differences and recognizing varieties". "Respecting differences and recognizing varieties" is the principle to be followed to achieve democracy, equality, fairness, justice and sincerity, and the basic guarantee of maximizing social vigor and mobilizing all factors conducive to social harmony.

In order to integrate cultural differences into an interdependent and harmonious cultural ecology, we need to identify and practice the concept of "respecting diversity and tolerating variety". "In this world, it is both right and necessary to recognize the same legitimacy of the basic premises that generate different cultural perspectives and value concept. It makes the mutual understanding and respect between different cultures possible:

---

[26] WRI, Guidiance on National Biodiversity Science Plan - Previous Experiences all over the World, assisted by the UNEP and the IUCN, translated D. Y. Xue, Etc., Chinese Environmental Sciences Press, Beijing, 1998.

This is the basic prerequisite to maintain peace and survive in a world with cultural diversity."[27] Cultures can only be promoted through exchanges in the social atmosphere of equality and mutual benefit, which is epitomized by harmony. As a multi-ethnic, multi-religious, multi-cultural nation, China still has a long way to go to achieve scientific development and build a harmonious society, but such endeavors and achievements are of great significance to the whole human society. In 2001, the 31st session of the General Conference of UNESCO adopted the "UNESCO Universal Declaration on Cultural Diversity", highlighting that "As a source of exchange, innovation and creativity, cultural diversity is as necessary for humankind as biodiversity is for nature."[28] As we all know, biodiversity is the basis of maintaining ecological balance. Why, then, cannot cultural diversity sustain our hope of world peace? In the process of building a harmonious society and promoting a harmonious world in China, we need to prove that cultural diversity is the basis of maintaining human peace. Modernization is a homogeneous quality of life instead of a homogeneous means of living. This is a challenging task facing all mankind[29].

Qinghai Province in China is a region with both biological and cultural diversities. At present, in the process of administrating and preserving the ecological environment of nature reserves, China has invested substantial money and technologies, and has identified definite goals. At the same time, Qinghai Province also took a series of effective measures, and the ecological environment problems of Sanjiangyuan, Hol Xil, and Qinghai Lake have been preliminarily relieved. As to the economic and social development of Qinghai Province, in the process of rapid expansion and acceleration of industrialization, great progress has also been made in energy saving and discharge reduction. However, there is still an increasingly arduous task related to human development and cultural diversity conservation, such as ecological migration. To solve this problem, we need to recognize cultural diversity as a valuable resource for the development of human society. Cultural and biological diversities are of equal significance to human society. The conservation of these "two kinds of resources" is the key to practicing the sustainable development of human-orientation

[27] E. Laszlo. The Destiny Choice: Survival Options for the 21st Century, translated by Y. B. Li, 1997, p.125.

[28] UNESCO Compilation of Documents on the Protection of Language and Cultural Diversity, compiled by J. J. Fan, 2006, p.100

[29] Zeng Z. Y. Fundamentality of Human Existence: Biological Diversity, Shanghai Science & Technology Press, Shanghai, 2002.

and the harmony between man and nature. This is precisely the point of view of eco-anthropology.

References

Brown L. R. Plan B, Ecological Economy, translated y Z. X. Lin, etc, the Eastern Publishing Company, Beijing, 2002.

Carrithers M. Why Humans Have Cultures - Explaining Anthropology and Social Diversity, translated by F. Chen, Liaoning Education Press, Oxford University Press 1998 version.

Chen J. S., Y. L. Cai, X. J. Wang. Human-Environmental System and the Sustainability, The Commercial Press, Beijing, 2001.

Clare M. T. War Resource: New Scene of Global Conflicts, translated by X. G. Tong and Zhi Ye, Shanghai Translating Publishing House, Shanghai, 2002.

Coleman D. A. Eco-Politics, Building a Green Society, translated by J. J. Mei, Shanghai Translation Publishing House, Shanghai, 2002.

Commoner B. Making Peace with the Planet, translated by X. L. Wang, Shanghai Translating Publishing House, Shanghai, 2002.

Council for International Cooperation on Environment, International Environmental Cooperation and Sustainable Development: Conserving China's Biodiversity, China Environmental Science Press, Beijing, 1997.

General Report Compile Group of China Biodiversity Conservation Action Plan, China Biodiversity Conservation Action Plan, China Science Press, Beijing, 1994.

Hao S.Y. Qinghai-Tibet Plateau Biodiversity Conservation and Environment – Taking the Qinghai Nature Reserve as an Example, Investigation Report, 2005.

Hardesty D. L. Ecological Anthropology, translated by F. Guo, H. Zou, etc., 2002.

Laszlo E. The Destiny Choice: Survival Options for the 21st Century, translated by Y. B. Li, 1997.

Morin E. & A. B. Kern. Terre-Patrie, translated by S. L. Ma, Joint Publishing, 1997.

Outline of Plan for Nature Reserves Development in China (1996-2010), approved by the State Council, State Environmental Protection Administration, State Planning Committee, issued on November 24, 1997.

Pearce D. W. and J. J. Warford. World without End: Economics, Environment and Sustainable Development, translated by S. Q. Zhang, China Financial and Economic Publishing House, Beijng, 1996.

State Environmental Protection Administration, China's Biodiversity: Country Study, China Environmental Science Press, Beijing, 1997.

Sustainable Development Research Group of China, CAS, China Sustainable Development Strategy Research Report 2007, Science Press, Beijing, 2007.

UNESCO Compilation of Documents on the Protection of Language and Cultural Diversity, compiled by Fanjunjun, Ethno Publishing House, Beijing, 2006.

Wang T. J. Population and Environment Carrying Capacity on Qinghai-Tibet Plateau, China Tibetology Research Publishing House, Beijng, 1998.

WRI, Guidiance on National Biodiversity Science Plan - Previous Experiences all over the World, assisted by the UNEP and the IUCN, translated D. Y. Xue, Etc., Chinese Environmental Sciences Press, Beijing, 1998.

Zeng Z. Y. Fundamentality of Human Existence: Biological Diversity, Shanghai Science & Technology Press, Shanghai, 2002.

Internet Sources

"An Important Sign for the Deterioration of the Ecological Environment: the Qinghai Lake becomes Salter", http://www.cws.net.cn/cwsnews/newshtm/qhh/.

"Ecological Environment Protection and Comprehensive Management Project of the Qinghai Lake Watershed Has been Started", http://news.sohu.com/20080526/n257087046.shtml.

"Endangered Species in Qinghai Province is Higher than the World Level", Qinghai News Net, June 27, 2005, http://www.sina.com.cn.

"Hol Xil Depopulated Area is facing the Dangerous Ecological Threat", June 28, 2005, Xinhuanet.com.

"Qinghai: China's Largest Ecological Migration Project Progressing Smoothly", http://www.ce.cn/xwzx/gnsz/gdxw/200710/05/.

"Qinghai Lake: the Anxious Gymnocypris Orzewalskii and Three Major Dangers", http://www.cws.net.cn/cwsnews/newshtm/qhh/.

"Qinghai Lake: Water Level Dropped Mainly due to Natural Disasters", http://www.cws.net.cn/cwsnews/newshtm/qhh/.

"Save the Qinghai Lake", http://www.cws.net.cn/cwsnews/newshtm/qhh/25.html.
http://www.qhepb.gov.cn/qhepb/index.asp.

Illustrations

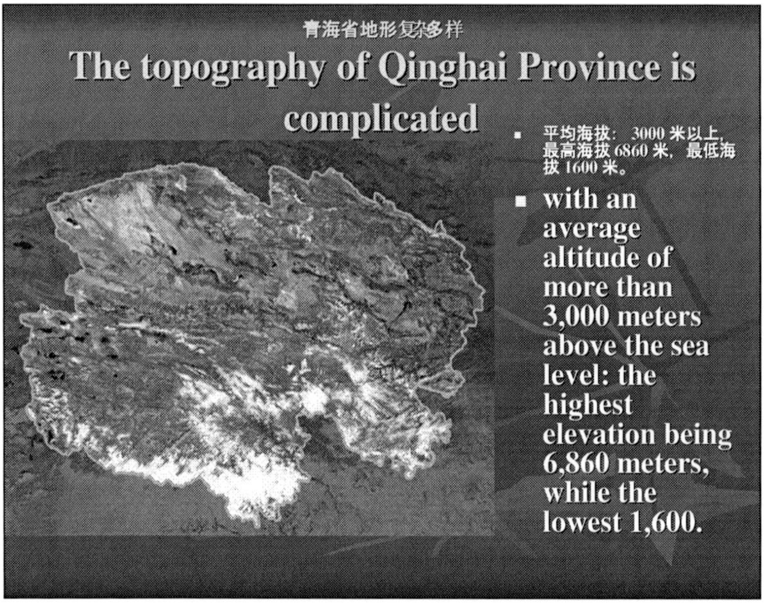

1. The topography of Qinghai Province

2. The Tibetan antelope (Pantholops hodgsoni), or chiru, is one of the most precious among the precious wild animals

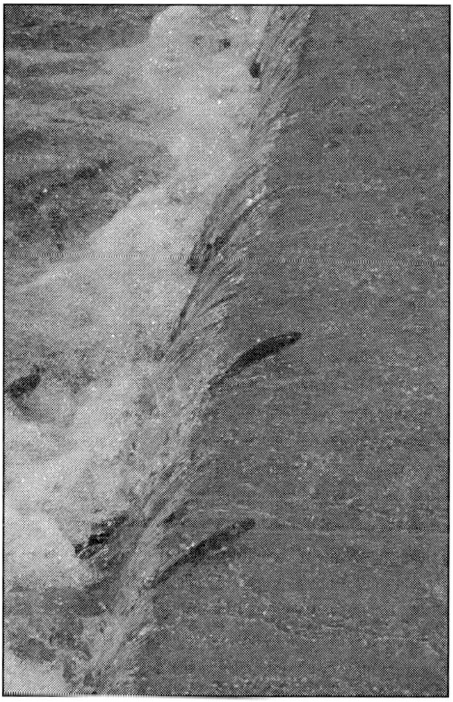

3. Between 2004 and 2006 a drastic increase has been seen in *huangyu* (Gymnocypris Orzewalskii) resources, from 5017.69 tons to 15274 tons

Abstract

Qinghai province is one of those regions in China which have specific physical geography and biodiversity, on the other hand there are also relatively heavy natural calamities there. There have been a series of ecological changes due to the reciprocal affects of natural and human elements during the process of the global warming-up and the local economic and social developments in recent years. Taking the two nature reserves Qinghai Lake and Hoh Xil as examples this article gives some analysis of the eco-environmental problems in Qinghai province including human activities (mainly referring to the economical ones). Then on this basis it discusses the harmonious developments of the population, society and nature from a point of view of ecological anthropology. It also introduces the progress of protection, rehabilitation and reconstruction of eco-environment in Qinghai with supports from the country, including achievements of natural disaster reductions.

Prof. Hao Shiyuan is a member of the Chinese Academy of Social Sciences and a Director of the Institute of Ethnology and Anthropology of the same Academy. Prof. Hao is also a Director of the Academic Division of Law, Social and Political Studies of the Chinese Academy of Social sciences, a President of the Chinese Ethnological Society and Chinese Society for Ethno-History, Vice-president of the Chinese Anthropological Society and the Chinese Society for Ethno-National Theory.

His scholarly interests are focused on ethnic problems and ethnic history, nation-state formation, nationalism and ethnic political development. He is an author of a great number of articles and books among which the most important are: *China's Nationalities and Ethnic Problems* (1996) and *Imperial Hegemony and the Balkan "Powder Keg"* (1999).

CULTURE OF ARID AREAS
A STUDY ON XINJIANG'S UYGUR PEOPLE

YANG SHENGMIN
PROFESSOR, DEAN OF SCHOOL OF ETHNOLOGY AND SOCIOLOGY,
MINZU UNIVERSITY OF CHINA

1. The Ethnic Groups & Population in Xinjiang

Uygurs are one of China's fifty-six ethnic groups. The Uygur population is more than 8 million, with over 90% of them living in Xinjiang Uygur Autonomous Region. Xinjiang is the largest province of China. The area covers one sixth of the entire country (Illustration 1).

As we know, energy resources are of worldwide concern. Xinjiang is rich in many kinds of mineral resources, especially oil reserves, which are the richest in China. Xingjiang has the potential to be among the main sources of such resources for China[1].

In addition to the Uygur, there are Kazak, Uzbek, Tajik, Hui, Tatar, Mongol, Xibo, Manchu, Russian , Daur, Kirgiz and Han people in Xinjiang. Most of them are Muslim and descended from Turkic ethnic groups (Illustration 2).

[1] J.T.Dreyer. The Xinjiang Uygur Autonomous Region at Thirty. *A Report Card in Asian Survey.* Vol.26, No7, July, 1986, p.721-744.

Table 1: Population of Ethnic Groups of Xinjiang (2000)[2]

Total	**17 915 459**		
Uygur	8 256 661	Russia	10 598
Han	7 023 910	Tajik	39 642
Kazak	1 277 474	Uzbek	13 198
Hui	813 023	Tatar	4695
Kirgiz	159 584	Manchu	22 329
Mongol	156 892	Daur	6405
Xibo	39 238	Others	91 810

The people of Xingjiang are closely related to the peoples of Central Asia in religion, history, language and culture, and they also have relationships with the peoples of the Middle East in the areas of religion and culture.

2. Common Characteristics of Different Groups in Arid Areas

I have been studying these ethnic groups, especially the Uygur, mostly from historical and anthropological approaches. To understand their history, I utilize archeological material and historical documents. To understand the present society, I depend on fieldwork and on-site investigations.

I have been engaged in studying these ethnic groups for about 20 years, and from the very beginning one aim of my study has been to inquire into

[2] *Xingjiang Statistical Yearbook of 2001*. China Statistics Press. Beijing, 2001.

and explain the cultural and historical characteristics influenced by the natural environment in which these groups live. It is my hope that this study will help us to understand the issues facing these ethnic groups today and in the future. It will also help us to understand other ethnic groups that have similar cultures and live in similar natural environments (Illustrations 3&4).

Here I mean the arid environment: encompassing 1.1 billion acres land, about 1/4 of the world is arid [3].

When I compare the ethnic groups in the Middle East with the Uygur people, I find that the cultures of different ethnic groups in arid areas have some common characteristics, as follows: 1. They are societies with centralized power; 2. Islam is usually the dominant religion, and often, mystical traditions have played very active public roles and 3. They highly value commerce and usually seek excellence in their commercial dealings.

3. Comparision between Uygur & Han People

In order to testify that the Uygur people value commerce more than farmwork, I had made an investigation in 5 villages in 1994. 2 of them are Han people's villages, another 3 are Uygur's. All the 5 villages are clustered together in small oases in the Turpan region of Xinjiang. Through the comparison of the different ethnic population in these villages, we can see that Han people are good farmers, and the Uygur people place a higher value on commerce.

Table 2: Comparison between Uygur and Han villagers on some economic targets (August 1994)

	Population	Farm land (acres)	Farm land (acres per person)	Deep wells	Output value (Renminbi per acre)	Agricultural Income (Renminbi per person)	Commercial Income (Renminbi per person)	Handiwork Income (Renminbi per person)
Han	513	1674	3.3	24	¥990	¥3231	¥12.4	¥49
Uygur	1432	2501	1.7	18	¥472	¥824	¥28	¥87

[3] K.Griffin. Rural Development in an Arid Region. *Xinjiang in Third World Quarterly,* Vol. 8, No 3, July 1986, p.978-1001.

Table 3: Comparison between Uygur and Han villagers on their Agriculture Labor & Inputs

	Times of irrigation (per a year)			Applied Fertilizer (kilogram per acre)			Time Spent on Farm-work (hours / day) (August)
	Soghum	Cotton	Wheat	Fertilizer	Urea	Multiple fertilizer	
Han	5	9	5	500	40	25	10---14
Uygur	3	5	3	250	10	0	8---11

4. Interpretations of the Scholars

In fact, some scholars such as Anne Marie Schemmel, Jo-Aun Gross, & Thomas Barfield have developed similar ideas through their research on the areas of Central Asia, the Middle East and so on[4]. Nevertheless, my theories are mainly based on the research on the Uygur in Xinjiang. So far in this field, no results of similar studies have been published by other scholars. Thus, I believe that my study may be useful in supplementing and modifying studies of other scholars.

In the early period of his research, Karl Marx developed the concept of an Asiatic mode of production, based on some ideas from classical economists, such as Adam Smith, James Mill, and Richard Jones.[5] After K.Marx, a similar point of view was held by Karl Wittfogel[6]. In fact, until now, some scholars still use the concept to explain the nature of oriental society. Thus

[4] J.A.Gross Muslims in Central Asia: expressions of identity and change, 1922, p.1-23; T. Barfield. Inner Asia and cycles of power in China's imperial history. – In: Rulers from the Steppe, ed. by G. Seaman, 1991, p.21-62.

[5] J. Mil. The Indian Form of Government – In: Orientalism, ed. by A.L.Macfie. New Haven, 2000, p.375-379.

[6] K.Wittfogel. Oriental Despotism: a comparative study of total power. New Haven, 1958.

the concept which was created 100 years ago by K.Marx might still have some relevance.

K.Marx argued that the agricultural base of oriental society is organized around large scale irrigation-works, possibly only under the sponsorship of a centralized government.

Consequently, autocracy and centralization of state power, became inevitable. In such a society, K.Marx proposed, the state is the superlative landlord, such that "private ownership" of land does not exist[7]. K.Marx was not able to complete his study on oriental society in his life time, but he appears to have believed that the Asiatic mode of production was applicable to oriental societies in general. My research, however, departs somewhat from K.Marx's view. In fact, I argue that the so-called Asiatic mode of production may have existed only in arid areas. Moreover, in these arid areas, the irrigation projects were not always built and managed by the government. The "Kareze systems", that are widely used in Central Asia, the Middle East and North Africa, have been mostly built by individual landowners. (In 1970's, there were 40 thousand kareze wells in Iran and 6 thousand in Afghanistan.) Based on this observation, the land is not completely run by the state but can be privately owned and managed.

Nevertheless, "centralization of state power" and "autocracy" is indeed an important characteristic of the arid areas. The research on the Turpan Basin shows us clearly why and how this preferred political structure was formed.

5. Environment & History of Uygur Culture

The Uygur people live mainly in the southern and eastern parts of Xinjiang, which constitutes the second largest desert in the world .The annual rainfall is less than 40 millimeters and the evaporation-rate is more than 2000 millimeters. There are few rivers, and resources-for-living are lacking in this area. It is one of the most arid areas in the world. Uygers traditionally settle around oases. These oases are like islands isolated from each other in the sea. Richard Bulliet once said the caravans that linked them were like ships moving between ports on the sea. The oases are spread along the edge of the Gobi desert. Some of the oases are very small, with only from one to a few dozen people living in them. Others are larger, with population ranging from several thousand to tens of thousands. The distance between the

[7] K. Marx. The British rule in India. – In: Orientalism, ed. by A.L.Macfie. New Haven, 2000, p.16-20.

nearby oases can be from ten to a few hundred kilometers. In some areas, several oases are clustered together, such as in Kashgar, Khotan, Kucha and Hami, and cities with populations of tens to hundreds of thousands have been formed. These towns and cities mostly are located near or at mountain passes where rivers run. Wherever there is water, there is an oasis; the bigger the river, the larger the oasis[8].

Humans appeared relatively late in these desert areas because of the severe environment. The earliest sites of human beings found in today's Xinjiang are only 6-7 thousand years old. During the course of several thousand years, peoples of different origins and ethnic groups came to Xinjiang from different directions. They spoke various languages and had distinctive cultures. The most important ancestral group of the Uygurs were the Huihe from Mongolia and the Saka from the grasslands of Central Asia. Over a period of several thousand years, while transforming the natural environment, they also transformed the social and political structure of their society and created a way of life that is suitable to the oasis. During this period, all of the cultures, experiences, and descendants of different ethnic groups interacted and mixed together. As a result, an ethnic group of the desert and the oasis, and its culture, were created – that is the Uygur culture.

6. Ancient Turpan Peoples – Nomadics

According to the general practices of anthropological field work, in order to investigate the genuine condition of the ethnic group and its culture, we must narrow the scope and try to probe into the internal communities of our research subjects. My research was concentrated on Turpan, the most arid basin in Xinjiang.

The area of Turpan is over 50 thousand square kilometers, with a population of more than 400 thousand, over 70% of whom are Uygurs. This area has provided the region's richest archaeological discoveries and historical documents. These discoveries and documents provide a long and detailed history of Turpan, since human beings first appeared there 6000 years ago and continue to live there today. Therefore, it is both typical of a Uygur community and sufficiently isolated to be intensively researched.

All early settlers of Xinjiang came from neighboring areas. They were mainly herders and hunters. Originally, they settled in the north of Xinjiang,

[8] K.Hickman. Oases for the Future in Science. *New Series*. Vol.154, No3749, Nov., 1966, p.612-617; J.Isaac. Oases in the Desert: Hannah Arendt on Democratic Politics. *The American Political Science Review*. Vol.88, No1, March, 1994, p.156-168.

in the Zungar grasslands and by 1500 BC, the increase of the population and the struggle for survival forced them to expand their settlement to the oases of south Xinjiang. Some of them also ended up in Turpan Basin.

The weather is cold and wet in north Xinjiang. With the vast grasslands being the major landscape feature, the land is best suited for animal husbandry. By contrast, south Xinjiang is hot and arid and the oases are quite narrow, thus, such a natural environment can only support small flocks of animals to be herded during colder seasons.

Therefore, the earliest settlers, the Che-Shi, only came into Turpan Basin to spend their winter at some warmer oases. These small oases formed their winter grazing land for the primarily nomadic people, and allowed them to settle for longer periods of time. Adobe houses were primarily built there. Eventually the winter residence became a permanent dwelling place which was different from the yurt on the grasslands. The elderly and children usually stayed there for the whole year. In order to gain access to drinking water, the winter residence was usually located near a river. Meanwhile, they cultivated small pieces of farmland on small plains by the river in order to make up for the decrease in herd size required by the smaller grazing areas[9].

After living in Turpan for approximately one thousand years, the Che-Shi people still relied on animal husbandry and hunting. Their agriculture was still small-scale and made almost no expansion. To establish and expand an irrigation agriculture at least two factors are needed: 1. A society that has a centralized administrative system, so that it can organize people to build large-scale irrigation projects, and 2. Compatible agricultural technology and means. The Che-Shi people didn't possess these two requirements, as they were traditionally a nomadic group.

7. Increase of the Population and Appearance of Agriculture

In the year 68 BC, the troops of the Han Dynasty (206 BCE–220 AD) occupied the Turpan Basin. To guarantee the provision of daily supplies to the army, the Han troops dug irrigation canals and cultivated farmland there. Therefore, Turpan was gradually developed into an agricultural area.

---

[9] J. Wilkens, M. Yaldiz (eds.). Turpan Revisited: The first century of research into the arts and cultures of the Silk Road. Berlin, 2004, p.49-53, 125-128, 138-140; J.Harmatta. History of civilizations of Central Asia; Vol.2. The development of sedentary and nomadic civilizations 700 BC, to AD, 250. Delhi, 1999, p.397-416.

The center where the Han army were stationed and which they cultivated was Gao-Chang city. The city was square, each side being 2 kilometers long. Around Gao-Chang city was the largest agriculture district in the basin[10].

Water was drawn from the mountain valley, through a 25 kilometers long main canal, then used to irrigate the farmland through the branch canals and ditches. Such types of large-scale agriculture, established and maintained by the army had not been possible under the organization of individual farming households. By this point, farm work was also completely undertaken by the army. It included complex, difficult work such as land reclamation, canal and ditch construction, allotment and control of water and so on. The system of collective labor and collective consumption was put into practice by the station troops. From then on, Gao-Chang city became the capital and the center of Turpan Basin until the city was destroyed during a war in 16th century.

After the Han Dynasty, more and more Han farmers who came from the central plains, migrated into Turpan. All the migrations were large in scale, collective in nature, and organized by the government.

After Han farmers settled down there, their social structure and system were transformed from their former ones when they were in the central plains. Since agriculture in central plains was mainly dependent on rainfall, it was possible to carry out the whole productive process by individual families. However, in Turpan, agriculture had to completely depend on artificial irrigation. Not only the farmland, but also all residents and animals depended on the water supply from the canal. Although the farmlands were distributed to individual families by the government, the construction and management of the canal, and the distribution of water supply was completely controlled by the government. Consequently, ownership of the land gradually shifted to the government.

Individual families only had the right of access. The state was the supreme owner. The result was a highly centralized mode of control. The government could decide and control virtually every aspect of the general livelihood of people in those areas.

---

[10] B. Thomas. Inner Asia and cycles of power in China's imperial history. - In: Rulers from the Steppe, ed. by Gary Seaman, Ethnographics Press, University of Southern California, 1991, p.21-62.

8. Characteristics of the Agriculture Society in Arid Areas

For over one thousand years, because of this mode of production, the social structure and system of the Turpan Basin has had the following four distinct characteristics: 1. Hereditary large family rule; 2. The lack of a formal distinction between the army and the people; 3. A combination of politics and religion, and 4. The nationalization of land.

I would like to precisely comment on two of these issues. Regarding the "**hereditary large family rule**": Because of its special geographical condition, Turpan had always been in a kind of independent or semi-independent position. All the immigrants of Turpan moved collectively in units of the whole tribe, village, town or county. After they settled, the former leader/chief of the tribe, community or the county would become a king with absolute power over the military, politics and religion. The main duty of the king was to manage the army and the canal systems. The king's son and relatives were assigned to different official positions in the army and the government. The last big family was the Uygur people. There were ten generations of Turpan kings who came from the same Uygur family. They ruled the Turpan Basin for over 200 years from 1731 till 1949. Except for the family of the king, usually there were several noble families which were related to the royal family through marriage, and therefore, also shared the same hereditary power with the king's family. They became high officials both in the central government and local government of the kingdom. Even nowadays, the distinction between the nobility and the common people is still noticeable among Uygur people, and this sense of differentiation can affect their intermarriage and social relationship with the association.

Regarding the "**nationalization of land**": There is basically no rainfall in Turpan Basin. More than 95% of the area is desert. Only 9 small rivers flow from the valley of the north mountain, creating some small oasis-segments. The water capacity of these rivers is very small. Moreover, the dry season lasts for 5 months every year. Irrigation and potable water all totally depend on these 9 rivers. All the rivers and their branch canals were controlled by the government. Since water belonged to the government, the land also belonged to the state. The officials of the government at all levels were the managers of all the different rivers, canals and ditches. So, the power of the government was highly centralized.

9. Kareze Well & Social Changes

Even though the water supply and cultivated land were limited, the population was continuously increasing. Consequently in the late Tang Dynasty (618-907), the Kareze well system was introduced. Since its introduction, the Kareze well system played a more and more important role in irrigation. Up until the 1960's, there were more than one thousand Kareze wells in Turpan. About 60% of the cultivated land was irrigated by Kareze well system. Apparently, the Kareze system had surpassed the rivers and the canals.

The Kareze system differs from the river and canal system in the following respects: 1. It draws upon the underground water flows all-year round with a stable water level; 2. Most Kareze wells were constructed with private investment and were not organized by the government. Usually, it only took 5-6 people to complete the whole construction of one such Kareze well. Each Kareze well could irrigate about 2 to 30 acres of land.

Schematic diagram of Kareze

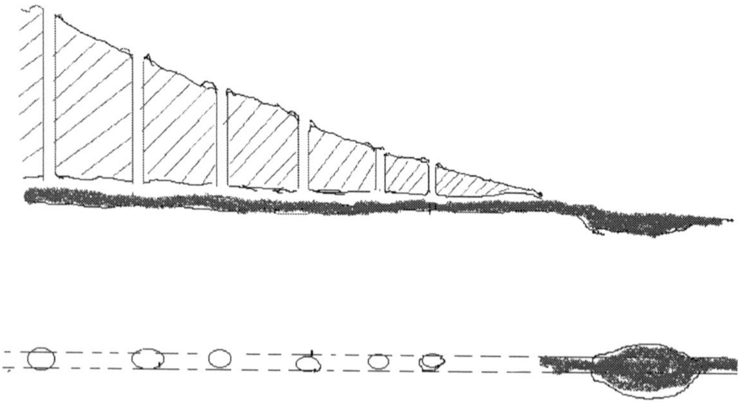

Because of this improvement in the irrigation system, around each of these Kareze wells, new villages began to appear in the desert. These villages did not rely on the state controlled river and canal system. The heads of these villages became the owners of the Kareze. Within these highly centralized societies, there appeared new villages, which became, to a great extent, economically independent and politically semi-autonomous.

Since the middle of the 9th century until 1949, Turpan was ruled by a succession of different kings. For over a thousand years, there were countless conflicts between the kings and the local chiefs. The kings controlled the river and canal system. But most of the Kareze system was managed and owned by the individual chiefs of the villages and districts.

As the river and canal system were much more powerful and more concentrated, they irrigated larger and more concentrated oases, including the towns that housed the county government. In contrast, the water flow in each Kareze well was much smaller, so it could only irrigate smaller and scattered settlements. The larger and the more concentrated system was clearly superior to the smaller, scattered one. Therefore, the kings always managed to maintain their superior rule. There are ample oral historical records from the old noble families about conflict between the Turpan kings and the hereditary local chiefs.

After the appearance of the Kareze system, though centralized power was diminished, the owner of the Kareze well became the master of the whole village. All the water and cultivated land on which all the villages rely, became properties of one family. Therefore, in these villages, another kind of social structure and relationship was established. All the villagers began to associate themselves with the Kareze owner, economically and personally. Each village was a small social community, autonomously controlled by the Kareze owner. Before the 1950's, most of the Uygur villages in Xinjiang were involved in such social system. A similar situation was also present in areas of Central Asia[11].

10. Conclusion

In my opinion, access to control over water, (or having the monopoly over water) is the fundamental reason for the presence and continuation of autocratic-power. In such society, all political and economic conflicts can be reduced to relationship that center around the control of the water system. It might be a bit difficult to fully comprehend such a phenomenon without in depth studies of other arid areas.

In conclusion, this study indicates that K.Marx was both right, and wrong. He was right that there is such a thing as an “oriental mode of production", but wrong to call it “orientalist”. Perhaps instead, we should rename it the “arid

[11] G.W.Anderson. Agriculture in the Undrained Basin of Asia. *Agriculture History*. Vol. 22, No4, Oct.1948, p.233-238.

mode of production". The comparative study of arid areas can teach us much about this mode of production and its influence on all aspects of life.

Based upon the data presented, the relationship between the arid environment and the Uygur culture can be seen in the following sketches:

Sketch 1. Arid & Culture

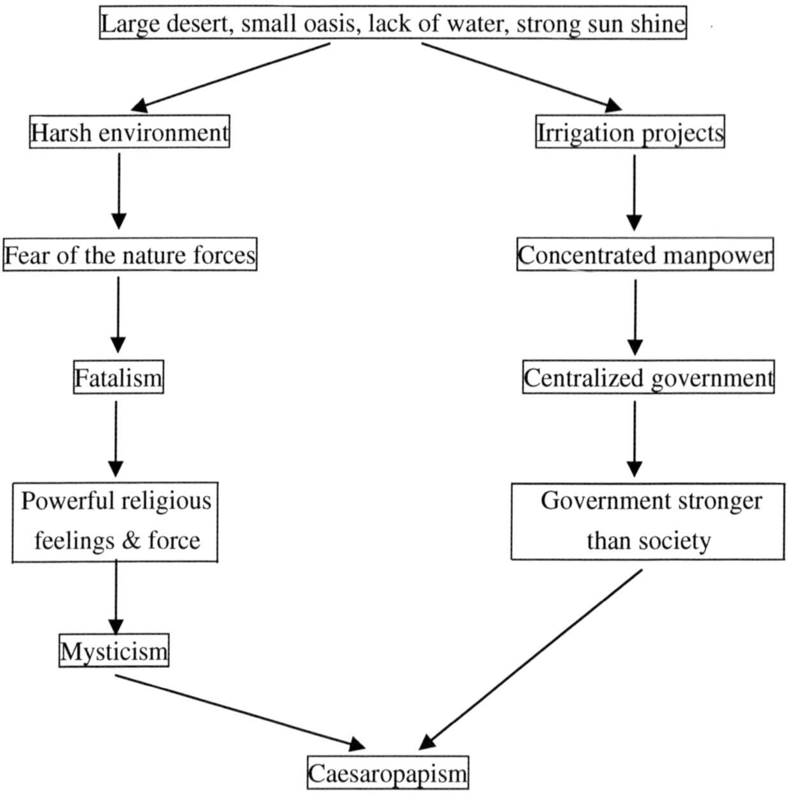

Sketch 2. Type of Economy

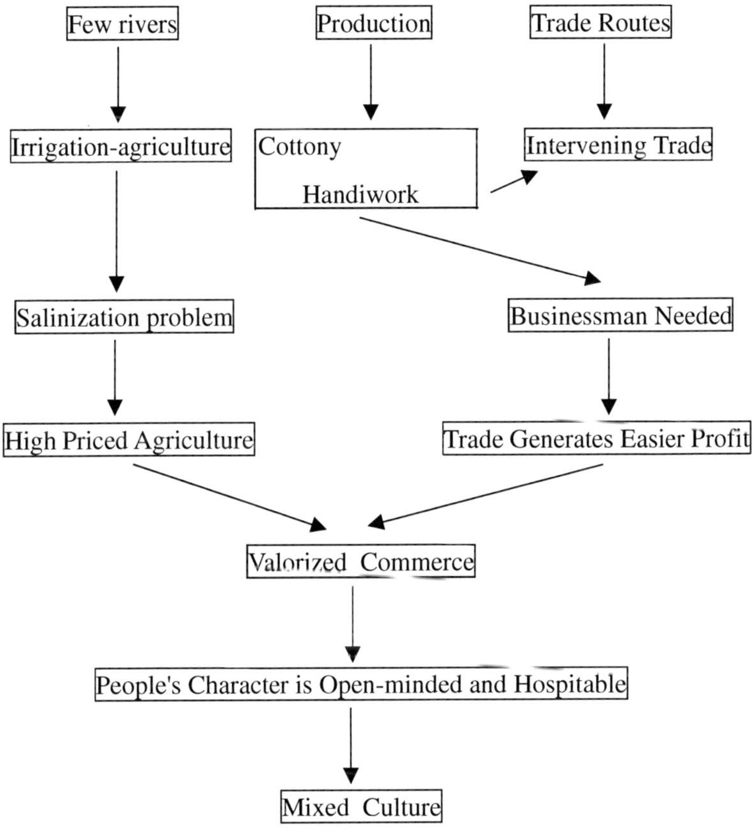

References

Anderson G.W. Agriculture in the Undrained Basin of Asia. *Agriculture History*. Vol. 22, No4. Oct.,1948, p.233-238.

Barfield T. Inner Asia and cycles of power in China's imperial history. – In: Rulers from the Steppe, ed. by G. Seaman, 1991, p. 21-62.

Dreyer J.T. The Xinjiang Uygur Autonomous Region at Thirty. *A Report Card in Asian Survey*. Vol.26, No7, July, 1986. Publ. by Univ. of California Press, p.721-744.

Griffin K. Rural Development in an Arid Region. *Xinjiang in Third World Quarterly,* Vol. 8, No 3, July 1986, p.978-1001.

Gross J.A Muslims in Central Asia: expressions of identity and change. Duke University Press, 1922.

Harmatta J. History of civilizations of Central Asia; Vol-2. The development of sedentary and nomadic civilizations 700 BC, to AD, 250. Delhi, 1999, Motilal Banarsidass Publ., p.397-416.

Hickman K. Oases for the Future in Science. *New Series*. Vol.154, No3749, Nov., 1966. Publ. by Assoc. for the Advancement of Science, p.612-617.

Isaac J. Oases in the Desert: Hannah Arendt on Democratic Politics. *The American Political Science Review.* Vol.88, No1, March,1994. Publ. by Amer. Political Science Association, p.156-168.

Marx K. The British rule in India. – In: Orientalism, ed. by A.L.Macfie, New York University Press , 2000, p.16-20.

Mil J. The Indian Form ofG. – In: Orientalism, ed. by A.L.Macfie, New York University, 2000, p.375-379.

Thomas B. Inner Asia and cycles of power in China's imperial history. – In: Rulers from the Steppe, ed. by Gary Seaman, Ethnographics Press, University of Southern California, 1991, p.21-62.

Wilkens J., M. Yaldiz (eds.). Turpan Revisited: The first century of research into the arts and cultures of the Silk Road, Berlin Reimer, 2004.

Wittfogel K. Oriental Despotism: a comparative study of total power. New Haven Yale University Press, 1958.

Xingjiang Statistical Yearbook of 2001. China Statistics Press, Beijing, 2001.

Illustrations

1. Location of Xinjiang Uygur Autonomous Region

2. The main population

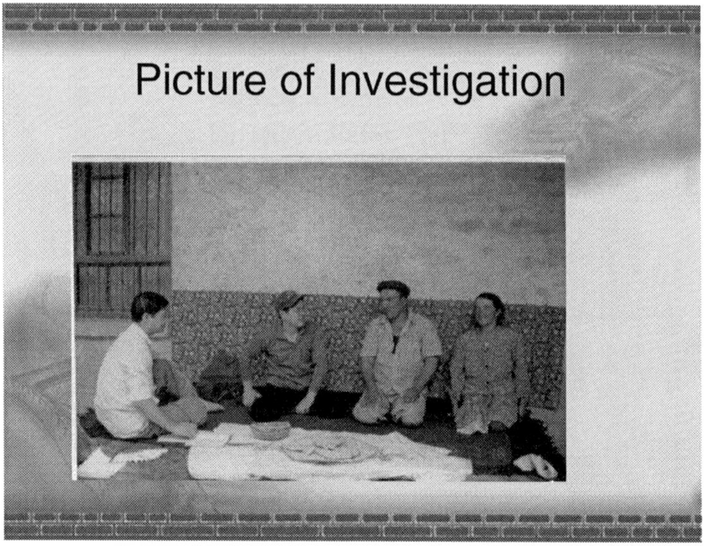

3. & 4. Among the Uygurs

Abstract

Uygurs are one of China's 56 ethnic-groups, living in Xinjiang Uygur Autonomous Region, and have been studied by the author during the last 20 years. For this particular subject he has examined 5 villages, 2 of which are Han villages, another 3 are Uygur. All 5 villages are clustered together in small oases in Turpan of Xinjiang. Uygurs have close relationships with the peoples of Central Asia in religion, history, language and culture, and they also have relationships with the peoples of the Middle East in the areas of religion and culture. The article investigates the common characteristics of the population in arid areas, makes a comparison between Uygur and Han people, examines the environment and history of Uygur culture as well as the ancient nomadic Turpan People. The conclusions made are represented in 2 sketches showing the relationship between the arid environment and the Uygur's culture, and the types of their economy.

Prof. Yang Shengmin, Ph.D. in Ethnology, Professor and Dean at the School of Ethnology and Sociology of the Minzu University of China. Major field of his interests are ecology and culture, culture in Arid and Semiarid areas, history and culture of ethnic groups in Northwest China and Central Asia. Among his publications are the monographs: *Uygur's History (1991); Notes and Textual Criticism on Historical Documents of the Turkish and Uygur people (1992); A Contemporary History of the Society and Politics of Xinjiang (1992); Ethnography of China (2003).*

THE CRISIS OF POST-SOCIALISM AND THE TRADITIONAL ATTITUDES OF THE BULGARIANS

RADOST IVANOVA

PROFESSOR, DSC, ETHNOGRAPHIC INSTITUTE AND MUSEUM, BULGARIAN ACADEMY OF SCIENCES

For Bulgaria, as well as for the other countries of the so-called "socialist camp" in Europe, the end of the 20th and the beginning of the 21st century was marked by significant changes which are only comparable to the most extreme events in history. The changes started with mass political protests in which the Bulgarians took a stand against the half-century long period of totalitarianism. They were swept off by the euphoria of the search for new perspectives, for new ways of life, the enthusiasm being followed, however, soon afterwards by hard times of frustration and desperation. In the 1990s, Bulgaria slid into a deep economic crisis, which shook the entire social life. In quite a number of people, especially in the elderly persons, this evoked nostalgia for "the orderly" life under socialism with all its scarcity, yet with its security. And, as it usually happens in turbulent times, the scum – that crime-generating layer which managed to seize money and power in muddy waters – floated to the surface. The majority of the common people either joined the army of unemployed or continued living barely above the level of subsistence with no prospects of better life. A great number of young people emigrated and are still emigrating to all parts of the world.

There are mass protests now as well, but they no longer have the characteristics of the clear-cut two-party opposition model – an opposition between the left-wing and right-wing parties – of the near past. When it is a matter of earning one's bread, of surviving in crisis, one's party affiliation remains in the background.

I shall not go into detail here describing the political and economic situation. My task, as an ethnologist, is to try and outline the changes taking place in the consciousness of the common Bulgarians, their way of thinking, their reactions and their approaches to the critical situations and developments. Because parallel to the process of socio-economic failure another process is also going on – the process of restructuring of the values, of accumulation of a powerful emotional charge which is inherent not only in the rhetoric of the meetings, but is also felt in many other spheres of life.[1]

The process of restructuring of the values in the post-socialist period rests, above all, on the cultural tradition inherited from the past, on the cultural memory. It is the phenomenon through which the experience gained by the previous generations is passed down to the next, and in which the prospects for the future are embedded. Although under the circumstances of significant changes the culture-determined codes of behavior, as well as of ethnic affinities and loyalties, rather show stability and can be qualified as "a second nature" of the individual,[2] essential changes can invariably be observed in them.

The new breakthrough in Bulgarian history marked by November 10th, 1989, resulted in a turnabout in the attitude of the Bulgarians towards their own identity. It may well be claimed that the prevailing trends developed on the basis of *a turnabout of values.* By renunciation of the hitherto hold cultural values the attention was directed to their opposites, most often found in the pre-socialist past. Moreover, the situation of mass protest movements created at the very beginning of the changes took the course of overturning the values to their extremes.

The crisis in policy and in society subjected to re-evaluation especially the attitude of the Bulgarians towards their daily lives, their workdays and holidays. They unambiguously rejected the national holidays of the Victory of the Socialist Revolution (September 9th, 1944), the Day of the Great October Socialist Revolution (November 7th) etc. pompously celebrated with parades and demonstrations. Instead, March 3rd, the day of the Liberation of Bulgaria from Ottoman occupation (1878) was declared a national holiday, along with other holidays of this kind. At the same time, as official holiday were reestablished Christmas, Easter, etc. traditionally celebrated prior to the Second World War which, under socialism, were not observed by the officials because of their associations with the Church. The renewed recognition of these holidays would not have been possible

---

1 M. Proshich-Dvornich. Modeli "retradicionalizacije": Put u buducnost vracanem u proslost. *Glasnik etnografskog instituta*, XLIV, Belgrad, 1995, p.303.

2 J.Roth. Blichwechel. Beitrage zur Kommunikation zwischen den Kulturen. Munchener Beitrage zur Interkulturellen Kommunikation. Band 15. Waxmann. Quoted after a book review by Troeva-Grigorova, E. *Bulgarska etnologia* [Bulgarian Ethnology], No3, 2004, p.89.

without the exoneration of religion. In fact, the return to religion became an act of "getting back to tradition, insofar as the generations of Bulgarians born and brought up in the totalitarian period used to interpret the religious faith as tradition".[3] On the other hand, via religion the Bulgarians, the overwhelming part of whom being Eastern Orthodox Christians, returned to the family of the big Christian community. What is more, the aggressive atheistic propaganda and the ban on religion under socialism have nowadays resulted in massive interest in the Church and its rituals. What is paradoxical in this case, however, is that the turn to the faith has become rather a fashion, because most of the church-goers are not familiar with the dogmas of the Church. It is precisely the freedom in this respect that shows the degree of alienation from the values of religion.[4] This notwithstanding, the Church now occupies a place of dignity in the holiday calendar and the life cycle of the Bulgarian people. Although the young couples contracting marriage, for instance, have no clear-cut attitude towards religion, they prefer getting married in Church, because the Church ritual gives the festive atmosphere so much looked for. Or the christening of newborn babies: secretly performed in the past, they now turn into festive events for the young families. We are witnessing christening of adult young men and women who cannot be married in Church, if they have not been baptized in Church before that.

Another evidence of the lack of knowledge of the postulates of the Eastern Orthodox religion is the large place that variable parapsychological phenomena and parapsychologists have come to occupy in the everyday life of the Bulgarians in the wake of the change.[5] The activities of the

---

[3] E. Krasteva-Blagoeva. Tsarkovniyat brak v Bulgaria sled 1989 g. – traditsia ili moda [The Church Wedding in Bulgaria in the Wake of 1989 – Tradition or Fashion]. – In: Obichaji zhivotnog tsiklusa [Rituals of the Life Cycle], Belgrade, Shtampa Rublishing House, Sofia, 2002, p.268.

[4] M.Benovska-Sabkova. Politicheski prehod I vsekidnevna kultura [Political Transition and Everyday Culture], Sofia. 2002, p.278.

[5] M.Benovska-Sabkova. Yavlenieto extrasensi ot gledna tochka na folkloristikata [The Phenomenon of Parapsychologists from the Point of View of Folklore Studies]. *Bulgarski folklor* [Bulgarian Folklore], No5, 1993, p.3-17; V.Ganeva-Raicheva. Paranormalni fenomeni i tyahnata proektsia v kulturata na bulgarina [Parapsychological Events and Their Reflection in the Culture of the Bulgarian People]. *Bulgarski folklor* [Bulgarian Folklore], No5, 1993, p.18-32; V.Dimov, L.Peicheva. Modernata magiya (Mezhdu "vsichko e magiya" I "takova neshto nyama" – opit za edin ne samo folkloristichen diskurs [Modern Magic (Between "Everything Is Magic" and "There Is No Such Thing": An Attempt at a Discourse, Which Is not Only Folkloristic). *Bulgarski folklor* [Bulgarian Folklore], No5, 1993, p.33-45; S.Stanoev. Ima li nyakoj tuk? [Is There Anyone In Here?]. *Bulgarski*

parapsychologists are a specific way of manifestation of the re-traditionalization of everyday culture, because they generally model on the activities of the traditional folk healers. On the other hand, this phenomenon testifies to the failure to distinguish between Eastern Orthodox Christianity and the belief in magic.[6] The return to religion and to the Eastern Orthodox Christian rituals, in particular, has nowadays yet one more function. It comes to fill the void left by the disappearance of the traditional rites, rituals and customs, which have already been forgotten. The temple is the institution which facilitates this process by offering its rituals ready-made, established by the canon. The role of the participants in them is comparatively limited and easy.

The change of the value orientation of the Bulgarians and their return to the traditions of the past, the revitalization of the cultural memory with the aim of overcoming the socio-economic crisis have acquired a special meaning in the political sphere. The frequent changes of governments formed of right-wing or left-wing parties' politicians in the 1990s did not result in a better life for the people. On the contrary, their living standards continued to deteriorate and they seemed to finally lose faith in those in power. Here I shall illustrate by one example the state of the political crisis at the turn of the new millennium and the efforts of the vast majority of people to overcome it following the cultural stereotypes of the past. This was the election, in 2001, of the former Bulgarian monarch, Tsar Simeon II, as a Prime Minister of the Republic of Bulgaria.

The election of Tsar Simeon II is one of the striking examples of the reorientation of the Bulgarians from tending towards the East – i.e. towards the former Soviet Union and the other countries of the socialist camp – to tending towards the West and the USA, admixed with a number of traditional views and beliefs of theirs. No doubt, of major importance here was the belief in the myth that the liberation of the Bulgarians had always been expected to come from outside, and with foreign support. Up to the beginning of the changes, Russia and the Soviet Union had been considered a double liberator and savior of the Bulgarian people: once, in 1878, from the Turkish occupation, and for a second time, in 1944, from the Nazism and fascism. Now the eyes of most Bulgarians were turned

---

folklor [Bulgarian Folklore], No5, 1993, p.46-55; I.Todorova–Pirgova. Bayaniya, magiy i seansi pri ekstrasens: ritualni vaplashtenia na predstavite za "dvata svyata" i vzaimootnosheniyata mezhdu tyah [Casting Spells, Magic and Sessions with Parapsychologists: Ritual Embodiments of the Conceptions of "The Two Worlds" and Their Interrelations]. *Bulgarski folklor* [Bulgarian Folklore], No5, 1993, p.56-68.

[6] M.Benovska-Sabkova. Politicheski prehod…, p.284.

towards the West, to Madrid, where living in exile was the former Bulgarian tsar. In this case, Simeon II was both foreign and own, in the same way as Russia and the Soviet Union had earlier been both foreign and own, with their state independence and their belonging to the Slav community. Simeon II is foreign, because he is a descendant of the Sax Cobourgs, their origins going to Great Britain; and he is own because he was born in Bulgaria, being the son of the last Bulgarian Tsar Boris III, who ruled up to the World War II, and an Eastern Orthodox Christian, like his father. Here, again, it is a matter of searching for solutions to the current and future problems looking back into the past and drawing on the traditional stereotype of being ruled by a tsar prior to the establishment of the republic in 1946.

At the elections in 2001, Tsar Simeon II was elected Prime Minister of the Republic of Bulgaria. His return from Madrid cannot be explained solely by a shift of the orientation towards the West. In my view, a great role was played by the patriarchal outlook, still preserved and not yet overcome, of the Bulgarians for whom the tsar was the good father from the fairy tales. According to the fairy tale ideology, kings and princes are of truly "blue blood". They are the secular representatives of the will of God on earth predestined to be endowed with supernatural abilities and virtues. The belief in these abilities was particularly strong in the past when the chief monarch – the tsar – performed important functions as head of state. Although the monarchy of the past used to have different functions from the monarchy of today, the modern popular conceptions of the monarchs do not substantially differ from their artistic embodiment in folklore.

Indeed, there is not a more suitable personality, closer to the image of the kings from the fairy tales than that of Simeon: born to a royal family in 1937; destined to succeed to the crown as the only son in the family; become an orphan at an early age (his father Tsar Boris III died in 1943); driven out of his Homeland together with his sister and his widowed mother in 1946; spent his life in exile, and so on and so forth. It is hard to find a character that could more aptly evoke compassion in the compassionate heart of the good Bulgarian, who has experienced and suffered a lot, than the real figure of Simeon as a small kid out of which the charismatic figure of the savior king could easily be created. And the political situation on the turn of the 21st century, as I have already mentioned above, with the two main political forces – the left `and the right – having been sufficiently compromised, was such as to pave the way for the re-traditionalizing of the mythical concepts of the supernatural abilities and virtues of the king. And a miracle did happen – Tsar Simeon II took the

Prime Minister post. The first words of the Prime Minister Simeon Sax Cobourg Gotha's, "Bulgaria is no longer the same", on the day of his election, seemed to be on the level of the miraculous, an element of the folklore poetry, and so was his address to the people, "Believe me, in 800 days Bulgaria will be completely changed." An analogy immediately comes to mind with Hans Christian Andersen's tale about the naïve king who liked so much to put on new clothes that he believed two cheats, who promised to make for him exceptionally fine clothes with the magic property of being invisible to those who were too foolish.

Obviously, the case of Simeon II shows that besides the belief in salvation from the West and, associated with it, the belief in the money, well-being and happiness, flowing thence to Bulgaria it is also a matter of the Bulgarian people's turning to tradition as a means of survival in a difficult situation. No doubt, this process of re-traditionalization in the political sphere cannot fully "repeat" the past by restoring the old political models – monarchy in this particular case. That is why in our case it was intricately intertwined with the process of modernization, which led to the unique figure of a king – prime minister. The restoration of the figure of the monarch in a modern role in a country with republican government reminds of the movement upwards the spiral which obligatorily includes going back. Indeed, it cannot be denied that every modernization puts into use the cultural models of the past. For Bulgaria, however, the processes of modernization have clearly been associated with laying particular stress on the phase of going back in a spiral backward movement to the past. And this return, by itself, slows down the development and provides conditions whereby the re-traditionalization is now extending also to such spheres of life as the government policy. And although by his life and behavior of a modern man Simeon II himself demythologized his own person, the belief in his charismatic features, charm, virtues, extraordinary abilities, etc. is still living in a great number of people and continues to function following the model of the fairy tale. As this example has shown, when the economic crisis continues too long, the re-traditionalizing of the monarchic past expands, too. The conception comes to life of the omniscient and omnipotent king who is expected to work miracles. For the time being, at least the first, and the only, miracle – a king prime minister – calls to mind the lessons taught by the folk classics, and the retort of the child from Andersen's fairy tale, "The King is naked" continues to be topical referring to simple truths.

The change of the government of Simeon Sax Cobourg Gotha in 2005, besides stripping the myth off the person of the tsar, showed the disappointment of most of the Bulgarians who were forced to forget about the fairy world of the tale and set their feet on firm ground again.

Although I have dwelt here just on limited spheres of the everyday culture of the Bulgarian people, I think that the examples quoted allow me to draw some conclusions about the use of tradition in situations which are critical for society.

What is common for the spheres discussed is that the revival of the cultural memory of the past is an important part of building the new cultural identity of the Bulgarian people. The restoration has, naturally, not been complete; it has been accompanied by some loss of memory made up for by the processes of modernization and globalization which take place not only in this country but all over the world, as well.

The cultural and historical experience of the past does not always contribute to the quick overcoming of the crisis. Sometimes it slows down the course of social development. Time is needed so that the delusions can be acknowledged. Vacillations and mistakes are the price the Bulgarians have been paying in order to get out of the crisis.

References

Benovska-Sabkova, M. Yavlenieto extrasensi ot gledna tochka na folkloristikata [The Phenomenon of Parapsychologists from the Point of View of Folklore Studies]. *Bulgarski folklor* [Bulgarian Folklore], 1993, No5, p.3-17.

—. Politicheski prehod I vsekidnevna kultura [Political Transition and Everyday Culture], Sofia, Prof. Marin Drinov Academic Publishing House, 2002.

Ganeva-Raicheva, V. Paranormalni fenomeni i tyahnata proektsia v kulturata na bulgarina [Para-psychological Events and Their Reflection in the Culture of the Bulgarian People]. *Bulgarski folklor* [Bulgarian Folklore], No5, 1993, p.18-32.

Dimov, V., L.Peicheva. Modernata magiya (Mezhdu "vsichko e magiya" I "takova neshto nyama" – opit za edin ne samo folkloristichen diskurs [Modern Magic (Between, "Everything Is Magic" and "There Is No Such a Thing": An Attempt at a Discourse Which Is not Only Folkloristic). *Bulgarski folklor* [Bulgarian Folklore], No5, 1993, p.33-45.

Krasteva-Blagoeva, E. Tsarkovniyat brak v Bulgaria sled 1989 g. – traditsia ili moda [The Church Wedding in Bulgaria in the Wake of 1989 – Tradition or Fashion]. – In: Obichaji zhivotnog tsiklusa [Rituals of the Life Cycle], Belgrade, Shtampa Rublishing House, 2002, p.267-278.

Proshich-Dvornich, M. Modeli "retradicionalizacije": Put u buducnost vracanem u proslost. *Glasnik etnografskog instituta*, XLIV, Belgrad, 1995.

Roth, J. Blichwechel. Beitrage zur Kommunikation zwischen den Kulturen. Munchener Beitrage zur Interkulturellen Kommunikation. 2003, Band 15. Waxmann. (Quoted after a book review by Troeva-Grigorova, E. 2004. *Bulgarska etnologia* [Bulgarian Ethnology], No3, p.86-89).

Stanoev, S. Ima li nyakoj tuk? [Is There Anyone In Here?]. *Bulgarski folklor* [Bulgarian Folklore], No5, 1993, p.46-55.

Todorova–Pirgova, I. Bayaniya, magiy i seansi pri ekstrasens: ritualni vaplashtenia na predstavite za "dvata svyata" i vzaimootnosheniyata mezhdu tyah [Casting Spells, Magic and Sessions with Parapsychologists: Ritual Embodiments of the Conceptions about "The Two Worlds" and Their Interrelations]. *Bulgarski folklor* [Bulgarian Folklore], 1993, No5, p. 56-68.

Abstract

The purpose of this paper is to outline the changes in Bulgaria at the end of the 20th and the beginning of the 21st century which took place in the consciousness and the way of thinking of the common Bulgarians, their reactions and their attitudes towards the critical situations and developments. The crisis in policy and society has subjected to re-evaluation the attitudes of the Bulgarians towards their daily lives, their workdays and holidays. The shift of values is illustrated in the paper mostly through the example of the rejection of atheism professed under socialism and the return to religion and Eastern Orthodox Christianity, the traditional religion of the Bulgarian people, in particular. Another example illustrates the state of political crisis at the beginning of the 21st century and the attempt of the people to overcome it pursuing cultural stereotypes from the past, namely, from the period prior to the World War II.

Prof. Radost Ivanova, DSc, has for many years been an Editor-in-Chief of the journal *Bulgarian Ethnology* and the annual *Ethnologia Bulgarica.* She is a Laureate of the European Herder Prize for 2002. Among her main publications are the books: *Bulgarian Wedding* (1984); *Traditional Bulgarian Wedding* (1987); *Epos-Rite-Myth* (1992); *Farewell Dinosaurs, Welcome Crocodiles! The Ethnology of Change* (1997); *Folklore of the Change. Folk Culture in Post-Socialist Bulgaria (1999); Culture of Crisis - Crisis of Culture* (2002); *Culture of Our Everyday life* (2006), etc.

SYSTEMS AND TECHNOLOGIES FOR SNOW DISASTER PREVENTION: ANTHROPOLOGICAL OBSERVATION ON THE PASTORAL AREA OF EASTERN QINGHAI-TIBET PLATEAU

ZHA LUO

PROFESSOR, INSTITUTE OF ETHNOLOGY AND ANTHROPOLOGY, CHINESE ACADEMY OF SOCIAL SCIENCES

The Planet's "third pole", the Qinghai-Tibet Plateau is the highest region in altitude. Due to the joint factors of high altitude and special meteorological conditions, snowfalls are a commonplace climatic component of the Qinghai-Tibet Plateau all the year round. However, in winter and spring, due to heaviness of the snowfall and extremely low temperature, the snow may remain for a very long time without melting. The pastures covered under a heavy coat of snow, it is difficult for the livestock to graze. Unfortunately, as a natural result of cold and hunger, many livestock die of weakness and sickness, causing tremendous property losses to the pastoralist households. For thousands of years, snow disaster has always been the uttermost threat to the production and life of the local herdsmen.

In order to prevent and mitigate losses caused by the snow disaster, the local pastoralists have taken an array of countermeasures such as the pertinent production arrangements, extensions of techniques designed for disaster prevention and mitigation and so on, attempting to maximally adapting themselves to the distinctive ecological conditions when it is still unlikely for human beings to intervene with the natural climate. In fact, after systematical inspection on the economic lifestyle of the nomads on the Plateau, one can easily discern their worries and concerns embodied therein toward disasters[1]. To some extent, it can be claimed that the unique economic lifestyle is built on these worries and concerns. However, this

[1] M.Goldstein and C.M.Beall. Nomads of Western Tibet: the Survival of a Way of Life. Hong Kong: Odyssey Productions Ltd., 1990.

does not mean that there are invariably constant and longstanding rules governing the systematic arrangements and technologies. Conversely, these systems and technologies are subject to constant changes, which are in turn dependent on climatic and environmental changes and affected by the constant evolutions of the human concepts on social development. What is of interest is that in various historical periods measures and strategies taken to address snow disaster threats have been nearly contradictory to each other. It must be noted that, despite unremitting efforts made, the local pastoralists are still by far very restricted in their capacities to effectively tackle the snow disaster risks. Many of them are highly exposed to poverty due to consequences of snow disaster losses, and have to rely on national and social relief and aid. Therefore, in order to foster the development of pastoralism on the Qinghai-Tibet Plateau, as well as to improve and boost the living standard of local herdsmen, it is necessary to constantly explore and improve the measures and strategies to help the pastoralists better prepared for snow disasters[2]. The initial step of this effort is to investigate the system changes over the past 100 years, as well as to analyse and assess various technical methods previously utilised.

Taken two pastoralist villages, the Mar Village and the Geri Village, in the eastern rangeland of the Qinghai-Tibet Plateau as cases for study, this paper presents a historical inspection and analysis of the systems and techniques for snow disaster prevention in the two villages over the past 100 years. The Mar Village is located in the northern Henan (rma-lho) Mongolian Autonomous County of Huangnan Tibetan Autonomous Prefecture in Qinghai Province. All the residents are Mongolians, who believe in Lamaist Buddhism, speak Tibetan and are fully assimilated to the Tibetan culture. Before the 1950s, the Mar Village was under the rule of the local hereditary Henan Mongolian Prince. The Geri Village is located in western Zeku (rtse-khog) County of Huangnan Tibetan Autonomous Prefecture in Qinghai Province. The villagers, historically part of the dpon-'khor tribe under the hereditary Hor stong skor (*qianhu*, or chiliarch), are all Tibetans. Both the two villages (the straight-line distance in between them is about 120 km) are located in the humid rangeland of the eastern Qinghai-Tibet Plateau and the upper reaches of the Yellow River. The forage here grows well in normal years, with the grass height in most areas of these two counties reaching 15 to 25 centimeters and on pastures with better temperature and moisture conditions even reaching 40 centimeters. This is in

---

2 X.Yu. Social Investigation on Nomadic Tibet in the Northwest of China. Beijing: The Commercial Press, 1948.

sharp contrast to the situation of the pastures in the central and western regions of the Qinghai-Tibet Plateau, where merely grows low and sparse forage. Yet, on the other hand, affected by the special geographical climatic conditions (here is the intersection zone where meet the warm and moist air currents from the south and cold air currents from the north), the two counties are where snowstorm most frequently hit[3]. According to related statistics, in the 38 years from 1956 to 1994, snow disasters occurred in a total of 15 years in southern pastoral areas of Qinghai Province, wherein includes the Henan and Zeku counties, with the chances reaching 39%[4]. Therefore, the two counties serve a good sampling for explorations into snow disasters.

1. Snow Cover and Low Temperature: Threats in Winter and Spring

Conventions and techniques for snow disasters prevention are disaster prevention and mitigation strategies established on the basis of long-term observations and accurate understanding of the features of local snow weather, along with local social and cultural conditions such as forms of social organizations and level of productivity (Illustration 1). Backed up by their forefathers' heritage for centuries of accumulated observations and experiences on the local climate, the herdsmen of the Mar Village and the Geri Village come to achieve a clear and definite understanding of the snowfalls and disaster factors.

It is clearly pointed out by the pastoralists that the incidence of snow disasters may well be attributed to thick snow cover and the extremely low temperature in its wake. Whenever heavy snowfall occurs, the pasture will be covered by thick snow, which prevents the livestock from grazing and eventually causes them to die of starvation. Related studies indicate that if the livestock are absolutely prevented from grazing, "the old, the weak and

---

[3] It is pointed out that there are generally three large regions of snow cover on the Qinghai-Tibet Plateau, but annual snow cover varies greatly in the eastern part, which is the major region suffering from extensive snow disasters. See: W. Dong et al. Analysis of Climatic Characters of Snow Disasters on Pastures on East Qinghai-Tibet Plateau. *Plateau Meteorology*, No4, 2001.

[4] W. Sun, and Y. Wu. Snow Disaster Sequence Chronology and Climatic Statistical Characteristics of Southern Qinghai Plateau. *Qinghai Meteorology*, No.1, 1996; K.Wen. Chinese Meteorological Disaster Dictionary: Qinghai Volume. Beijing: Meteorological Press, 2007.

the sick will die on the fifth day, the average livestock on the seventh day, and on the tenth day the majority of them will die."[5] On the other hand, the harms of extreme cold are twofold: it prevents the snow from melting, while at the same time directly threatens the survival of the livestock. According to the pastoralists' observation and experiences, the livestock lose weight rapidly in low temperature conditions, and get sick easily. Modern science indicates that under low temperature conditions, livestock have to decompose the body fat to keep the temperature balance, and severe cold may cause the excessive consumption of body heat of the livestock, which leads to rapid decline in fat reserve of the livestock. As a result, those thin and low-weighted livestock will be debilitated and even die within a couple of days.

Moreover, the special timing of snowfalls in those localities also contributes to the frequent incidences of snow disasters. According to the pastoralist respondents, the snowfalls occur more frequently in winter and spring on the Qinghai-Tibet Plateau; however, unlike the other places, what is a distinct feature in the local climate of the eastern regions of the Plateau is that "it snows early, and sometimes snow disasters even occur as early as late autumn, but the snowing season ends very late, and severe snow disasters often occur in early summer." Snowstorms frequently hit in late autumn and early winter, which signifies that the pastoralists must move back in good time to the winter pastures before wintertime, and it will be very hazardous to stay in the summer and autumn pastures. If the snowfall in early winter is followed by drastic temperature drop immediately in its wake, it is undoubtedly the beginning of a nightmare to the pastoralists because it will be getting colder and colder and snow can not melt for a long period. Consequently, the livestock may successively die of hunger and cold. Compared with the winter, spring is more risky because there are more snowstorms in spring than in winter[6]. To make things even

---

[5] This is a conclusion drawn in accordance with the research in Inner Mongolian pastoral area (See: D. Gong, M. Hao. Synthetical Indexes of the Disaster of Snow Cover. *Journal of Applied Meteorological Science*, September 1998); the investigation result of Qinghai Province is nearly the same (See: L. Zhou. The Basic Characteristics of Heavy Snowstorm Process and Snow Disaster Distribution in Eastern Pastoral Areas of Qinghai-Tibet Plateau. *Plateau Meteorology*, No4, 2000.

[6] According to the statistics on the number of days of snowfall in Henan County from 1959 to 1980, there are 53 days with snowfall in Henan County on average per year, 34.6 days of snowfall is from January to May, accounting for 65% in the whole year. *Henan County Annals*, by the Local Records Compilation Committee of Henan Mongolian Autonomous County. Lanzhou, 1996, p.184.

worse, the conditions on the pastures in spring are greatly worsened. After a whole winter's grazing, the withered forage has been eaten to the roots; even if there is withered grass to feed, its nutritional value is extremely low,[7] and "the herd and flock only get semi-full even if they graze a whole day, and there is no nutrition at all." The livestock only have skin and bone after a winter, with poor resistance to diseases. At this time, "One spring snowfall, covering the grass for only two or three days, can mean the death of scores, and a spring blizzard can leave the pastoralist with no herds at all"[8]. Spring is the most critical season for pastoralists.

Due to the climate features of frequent heavy snowfalls in early winter and late spring, the pastoralists are obliged to stay for at least 6 to 8 months on the winter and spring pastures, which will inevitably lead to shortage of forage as a result of the herds and flocks' long-time stay.

How to get prepared for the snowfalls and low temperature in winter and spring seasons, or more precisely, how to address the scarcity of livestock feed after snow and how to remedy the injuries inflicted on the livestock by temperature drop following snowfalls are key to prevent snow disasters. The system design and technical strategies of the Mar and Geri villages for snow disasters prevention have been revolving around these two issues. However, solutions worked out in different eras are entirely different.

2. Traditional Tribal Era: Taking Full Advantage of the Natural Resources

In traditional Tibetan nomadic communities, the tribes were both the basic political units and economic organisms (tribal members shared and possessed the fixed pastures). Generally, a tribe was composed of a certain number of "ru-skor", which served as grass-root organizations of economic production and social mutual aid. Although tribal members grazed the herds privately owned by their respective households, the pastures, the most important means of production, was a collective asset of all tribal

---

[7] Related research shows that crude protein content of forage in grass withering period (from November to the following April) is less than half of that in summer: N.Qin. Rebuilding Historical Climate Data and Research on Climate Changes in Qinghai Province. Beijing: Meteorological Press, 2006.

[8] R.B.Ekvall. Fields on the Hoof: Nexus of Tibetan Nomadic Pastoralism. New York, 1968, p.31.

members (there were isolated cases where the tribal heads might own their private pastures, though). Therefore, the economic activities of traditional nomadic tribes were characterized on the one hand by household grazing and private ownership of herds and on the other hand by organizational operation and collective ownership of pasture resources to some extent.

A small tribe under the jurisdiction of a Mongolian Prince at that time, the Mar Village community was divided into five "ru-skor". They divided the pastoral areas into three seasonal pastures: the winter and spring pasture, the summer pasture and the transitional pasture in between the former two (known as the autumn pasture, though, it was actually exploited in spring as well). Every year in late April and early May on the Tibetan calendar when the weather got warm and the grass turned green, they left the winter pasture where they had spent more than six months, and moved about 30 to 60 km northward to the transitional pasture, where they would spend another two months. In early July, again they would move 20 km further northward to the summer pasture. Two months later, they would return to the transitional pasture once again. Around September 20, they would again return to the winter pasture, where they would remain until the pasture germinated in the next year. The migration pattern of the Geri Village in most part paralleled that of the Mar Village, only that the seasonal pastures of the Geri Village varied greatly in elevation. Therefore they were more conscious to the seasonal changes in climate. This perennial cycle pattern of seasonal migration was not to seek better pastures; on the contrary, almost all the summer pastures were meadows relatively more arctic-alpine, where the conditions were more difficult. Rather indeed, the pastoralist migration of the Mar and the Geri villages might well be understood as makeshift strategic movement away from the winter pasture mainly with a view to get prepared for the snow disasters and safeguard the economic security during the snow seasons.

As already mentioned above, one of the major causes for the snow disasters lies in thick snow cover of the pastures, thus preventing the herd from grazing. The most common solution for dearth of forage is to reserve forage or fodder in advance. However, the nomadic pastoralists on the Qinghai-Tibet Plateau generally did not have the conventional practices of storing fodder in that the pastures on the plateau were generally sparse and short forage made difficult mowing and collecting. Although the pastures of the Mar Village and the Geri Village were lush, the pastoralists did not have the convention of collecting grass for store. It is maintained that the herds then were too small and such labor was not worthwhile. However, considering the high mortality rates after snow disasters, this point of view is clearly not convincing. R.Ekvall proposes an alternative interpretation regarding the absence of such conventions that forage mowing would lead to primitive claiming of household land exploitation, which was in conflict with the tribal concept of communal pasture and grazing[9]. It is very likely that just out of the consideration of preserving the public ownership of pastures, which was more significant to the tribal members that restrained the mowing activities of the pastoralists.

Thus, the pastoralists turned to another alternative, that is, to avoid the snow disasters they tried to foster the height of natural forage so as to prevent from snow covering. Experiences and observations told the herdsmen that it was directly related to the height of forage whether a certain amount of snow cover would result in disasters. In the area where the forage was tall, the tips of grass would protrude above the snow cover. As long as they could graze, even merely on the tips of forage, the livestock could survive, whereas it was fatal when there was nothing to graze at all. Just based on such observations, the pastoralists were able to reasonably arrange their pastures according to seasonal shifts: those pastures in warm areas with excellent vegetation were exploited as winter pastures, while those in the areas of higher elevation and harsher weather conditions – where, though, in summer there was usually enough forage to graze – as the transitional or summer pastures. Hence their pasture utilization system, viz., the strict nomadic system, was to set aside for the snow seasons a fully-grown pasture with possibly maximal grass height, where it was kept from grazing in summer. During the vegetative periods, the pastoralists left the winter and spring pastures and moved to and grazed on other pastures to keep clear from the winter and spring pastures and help foster the forage growth there. In this way, when winter came, they could not only have a "naturally" fully-grown pasture with a substantial store of forage, but could, by virtue of the natural height of grass, minimize the risks of the pastures from being deeply covered underneath snowfalls.

In order to fulfill the above aims, the collectivism property of the pastures was attached great importance by the pastoralists in their exploitation. Guidelines for collective actions and related rules and regulations were established to ensure the smooth working of the whole process of seasonal migration and the pasture utilization system.

First of all, the tribal members followed the principle of unified action during seasonal migration. In general, the winter and spring pastures were low in altitude, warm in climate and more comfortable in living conditions;

---

[9] R.B.Ekvall. Op.cit., p.35.

therefore, there were always some members who expected to postpone the migration away from them and return as early as possible, especially the elderly and children, who were even more reluctant to leave the winter and spring pastures. However, to leave some people on the winter pasture meant that a necessary number of livestock also had to be left behind, such as horses for journey, and yaks for milk, which would certainly affect the restoration of the winter pasture to some extent. What was even worse for this practice was that it might result in the possible domino effect and more members would follow their examples and join in this practice. In order to address these problems, both the Mar and the Geri villages had the collective consultative system, or pastoralist representative consultative system to determine such issues as the distribution of pastures, migration timing, and whether to leave people on the winter pastures. In the Mar Village, these issues were negotiated by five elected headmen of the ru-skor, while in the Geri Village they were decided by "khyigs-bdag" – a specialized team composed of pastoralist representatives from the four ru-skor under the dpon-'khor tribe. When it was time for migration, all members were required to move within the specified time (2 to 3 days), with no delay permitted. The Mongolian Prince would send pasture management personnel to inspect the winter pasture of the Mar Village. If any livestock (which belonged to the herders still staying on the winter pasture) was spotted after May 4th on the Tibetan calendar, the rule-violating pastoralist involved would be punished by having his herd slaughtered.

Secondly, the pasture inspection system was practiced. Under tribalism the various tribes were politically independent. Therefore, the incidence of mutual pasture raiding and seizing was commonplace, and cross-border grazing on pastures of other tribes occurred from time to time. In order to ensure full recuperation of their own pastures, the tribes practiced the regular conventional pasture inspection systems, paroling the farthest ends of their pastures and expelling the trespassing pastoralists and herds. More tribal members would be recruited into the paroling team during the eventful periods, such as when a tribe fell into dispute with a neighboring tribe over the pastures. It was necessary that each and every household present one male adult member to the patrolling team; those that could not provide a member had to pay in compensation for their absence.

During the traditional tribalism era, two main methods were employed by the pastoralists to address the hazards of severe cold. The first was to make every possible effort to implement the warm keeping measures to protect the livestock. For example, the winter settlements were established on the leeward slopes or in the coves, which exposed to the sun.

The pastoralists of the Mar Village were not accustomed to building livestock sheds, not only because of the lack of building timbers on the pastures, but also because of the pastoralists' lack of civil engineering technology, and more importantly, because of the herders' lack of incentive for investing in shed construction because to embody the fairness of pasture utilization, upon each seasonal moving, the camp sites of all *ru-skor* were determined by drawing lots; therefore, the annual locations of "ru-skor" were different. They simply used clods or turf to build walls to serve as livestock barns to ward off the cold wind. However, this type of livestock sheds were in the open air and thus only had limited effect in resisting the low temperature; as a result, many livestock got sick or even died of cold at night. The other method is to improve and maintain the fat reserve and weight of the livestock through carefully shepherding so as to help them in possession of desirable cold-resistant body conditions. The diligent herders developed a range of livestock shepherding techniques, which were sufficiently effective to ensure the best fat reserve and weight for the livestock when winter came and to help them survive the harsh winter. In fact, the extent of herd loss in the disaster was always closely associated in proportion with the diligent efforts and experiences of the pastoralists. Hence, the number of livestock died in the snow seasons was often held as a criterion to judge whether a herder was diligent or not.

In short, the collective possession of pasture resources was conducive to the unified management of pastures. Collective decision-making within the grass-root organizations concerning the pasture utilization rules and collective supervision prevented the arbitrary disturbance of the winter pasture. The private ownership maximally enhanced the pastoralists' enthusiasm for keeping the livestock. Hence, this system was designed to take full advantage of the natural resources to prevent snow disaster. For the pastoralists of the Mar and the Geri villages, so long as they strictly controlled every step of the system, the above strategies might be integrated to bring about anticipated effect on disaster prevention. However, the herders were very clear about the obvious limitations inherent in this disaster prevention system. As neither were they traditionally used to storing forage for times of dearth, nor was there any possibility for them to obtain external assistance, such as the government's relief, they were completely helpless when snow cover and temperature drop reached a certain degree. In the case of the winter of 1954, it snowed continually for more than thirty days, and the average snow cover reached 30 to 50 centimeters. Concomitant with it were strong wind and drastic temperature drop. The pastures were snow-covered for days on end. At that time the newly established government of Henan County did not have sufficient capacity to implement relief intervention. As a result, a large number of the livestock were frozen to death or starved to death, with the mortality rate as high as 50%[10].

3. People's Commune Era: Introduction of Disaster Prevention Technology

Five years after the founding of the People's Republic of China in 1949, the People's Governments of the Zeku and Henan counties were established. The establishment of the new regime gradually put an end to the termination of the traditional tribalism and established the new socialist system step by step. After nearly 4 years' (from1954 to 1958) transitional socialist reforms of Mutual Aid Team and Production Cooperative, the Mar Village and the Geri Village were transformed into an integral part of the *Renmin Gongshe* (the People's Commune, or collectivism) in 1958, and the previous tribes and *ru-skor* were replaced by *gongshe*'s or *shengchandui*'s (production teams). The new system was foremost characterised by the public ownership of the means of production, that is, both the pastures and all the livestock previously owned by individual households turned to be collectivist assets (in later stage of the collectivism system private possession of a small number of the livestock was permissible); the production was no longer conducted by individual families, and the families were no longer the unit for accounting. On the other hand, unified planning regarding production and accounting was made exclusively by the collectivism and the production team, and the herders were supposed to complete the tasks assigned by the production team, all labor being transformed into *gongfen*'s (quantitative indicators) and noted down in a *gongfen* album. At the end of the year the amount of *gongfen*'s determined a herder's personal income, which was calculated on the basis of the collective income. This form of ownership lasted for 27 years in Henan County and Zeku County until the implementation of the Household Responsibility System reform in 1984.

Under the new system, although the form of ownership changed, the mode of production was still a continuation of traditional nomadic pastoralism, with the pastoralists still grazing on the original pastures, practicing

---

[10] *Henan County Annals*, Op.cit., p.185; G. Shi Natural Disasters in Qinghai. Xining: Qinghai People's Press, 2003.

the inherited traditional techniques for disaster prevention. However, in the face of the long-term shortage of products, the new local government officials believed that the high mortality rates of the livestock in winter and spring were unacceptable, and expected to take more initiatives to mitigate disaster losses. The proactive strategies taken may be generalized into the following two aspects: the first was to adopt the effective techniques by way of renovating and integrating those traditionally used by the pastoralist predecessors and promote their extensions via institutional efforts; the second was to actively introduce external technologies, including mainly wasteland cultivation and forage planting as well as the construction of the "enclosed-pastures with fences" (Illustration 2).

3.1. Extensions of Traditional Techniques: Construction of Stockyards and Livestock Sheds

In the traditional society, although all people realized that the construction of stockyards and livestock sheds might improve temperature conditions at night and in winter, which was conducive to the disaster resistant capacity of herds, only the tribal heads and a few well-off households were capable of building good livestock shelters. During the collectivism period, the collective possession of the livestock and the system of grazing the livestock in separate herds and flocks resulted in labor surplus in the production team. In traditional society, at least two labors were needed for each family to graze the herd and the flock separately; in fact, some households had such a small herd that the workload was very light. After the collective integration, the scale of the herd and flock were enlarged (generally speaking, 200 to 250 ewes forming one group, 400 to 500 wethers and one-year-old lambs forming one group, 40 to 50 horses forming a group, 80 to 100 two- to three-year-old calves forming a group, 60 to 75 cows forming a group, and 50 to 60 remaining calves forming a group[11]), whereby labor surplus emerged, who could be involved in the two aforementioned activities, that is, construction of stockyards and livestock sheds as well as wasteland reclamation and vegetation cultivation. In addition, supported by the governments' policies, the herders were allowed to cut timbers in the adjacent forest areas for the construction of livestock

[11] Anonymous Man's Method for Implementing the Principle of Distribution According to Labor Done (Draft), May 12, 1961. *Zeku County Annals*, by the Local Records Compilation Committee of the Zeku County. Beijing, 2005, p.564.

shelters, and sometimes they were furnished with raw materials. The extended construction of stockyards and livestock sheds undoubtedly had significant effect on the livestock for their survival in the winter and their resistance to severe cold after the snow (Illustration 3).

3.2. Introduction of New Technologies: Wasteland Cultivation and Vegetation Planting, as well as the Construction of "Enclosed-pastures"

Since the core problem with the "natural death" in winter and spring seasons or death from snow disaster was the dearth of forage and lack of temperature-keeping measures, it should be solved through targeted measures. The common agricultural practices such as plant cultivation, and forage collecting and storing were introduced and extended first. The introduction of agricultural technologies was associated with the family background of local administrative personnel. In order to complete the reforms of local society and establish a new political and economic system, a number of administrative staff was sent to this area. Most of these mission-bound cadres were either of the Han Ethnic Group from inland China or of the Tibetan Ethnic Group from agricultural areas of Eastern Qinghai Province, and their rural background entailed on them a natural preference for the mode of production prevailing in the agricultural community. To them, the pastoralists were too non-proactive in addressing snow disasters, so they proposed that positive measures such as wasteland cultivation and vegetation planting, as well as forage collecting and storing should be taken to prevent snow disasters. Under the guidance of this outlook, the pastoralists' patterns of disaster preparedness and mitigation, and even their life were subject to a series of changes.

Looking back, rgya-mtsho, a pastoralist from the Mar Village recalled that: "From around 1958 onwards, the Government mobilized the members of the communes and production teams to plant grass and develop agriculture. Farms were set up on the banks of the Yellow River, and members from my village were dispatched to farm. As it was too cold to grow grain successfully, the farms had to be abandoned. However, such activities as forage collecting, and building livestock shed and stockyards continued. An area of 80 mu (about 5.36 hectares) in my village were reclaimed as farm land for oat planting. At that time, we had to spend one month every autumn collecting grass. However, most of the forage prepared for snow disasters was fed to the cadres' horses, and only a very small part was left for the livestock's winter feed. Hence grass collecting did not help much in disaster resistance. The construction of livestock sheds and stockyards were common activities then. The team leader annually checked the livestock sheds and stockyards, and arranged labors to repair damaged ones. In order to improve the pastures, we also constructed the water conservancy projects at the Government's call, such as drilling wells, water diversion, and water channel building, and most of the materials and equipment were provided by the Government. Water conservancy projects facilitated obtaining of drinking water both for the people and the livestock, but people paid no attention to protect them at that time, and when the water channels were damaged, nobody cared to repair them. By and by, those water conservancy facilities were abandoned. In short, we were pastoralists, but in those years we often worked with the spades and sickles, just as the agriculturalists."

The introduced new technologies were theoretically helpful to improve the capability of disaster prevention and resistance, but they actually caused damage to the pastures as the local natural conditions were neglected during their implementation, and many non-arable pastures were opened up. According to a piece of official data, "from 1958 to 1960, a total of 572 million mu of land were reclaimed in pastoralist areas (in Qinghai Province); the even and good winter pastures within the six pastoralist counties were cultivated, and more of the weak and thin livestock died of severe dearth of forage in winter and spring.[12] "In Henan County, a total of more than 20 million mu of rangeland was reclaimed from 1959 to 1962, accounting for 2.02% of the available pastures in the county. Large areas of fine pastures are destroyed, which have not been restored even when cultivation were stopped 20 years ago"[13]. However, the introduced new technologies have achieved a certain degree of success on both banks of the *ba-chu* River near the Geri Village in western Zeku County, which are lowest in elevation in Zeku and enjoy good irrigation conditions. Cultivated vegetation (including the annual and perennial types) has greatly increased unit forage yield and facilitated forage harvest and storage, which significantly improved local herders' resilience to disasters. Therefore, later, when the mistake of pastures reclamation was corrected, the reclamation and vegetation planting in that place are not stopped but rather encouraged instead. With the gradual expanding of vegetation cul-

[12] H. Niu. Historical Data on Cooperative Economy in Qinghai Agriculturalist and Pastoralist Areas. Lanzhou, 1993, p.283.

[13] *Henan County Annals*. Op.cit., p.266.

tivation area, the Geri Village and other neighboring villages have benefited, their capacity to resist snow disaster being significantly improved.

The pastoralists have complicated feelings regarding these new technologies. The practice of building livestock sheds and stockyards construction derives from their past production experiences, and forage reserve is clearly helpful to ease the shortage of livestock feed after snow. However, in the memory of the pastoralists, the body conditions of the livestock of the production team were not desirable, or even worse than in the past, and the mortality rate in spring and winter remained high. All this was ascribed to the public ownership of the means of production and egalitarian distribution principle. Under such systems, the pastoralists' efforts of protecting pastures and managing the livestock did not have much to do with their income. Therefore, they lacked the enthusiasm in the operation and management of the pastures and the livestock. A pastoralist named 'jigs-med observed that, "due to the collective ownership, the collectivism was also supposed to take the losses, which had little to do with individual households. People did not have a strong sense of responsibility for reducing losses caused by snow disasters. Traditionally, the pastoralists would first move to the pastures with poor natural conditions, and finally to good ones, so as to leave something for untimely needs. However, in the period of collectivism, the arrangements for the pastures were in a mess, and people went to graze wherever the forage was lush. As long as they finished the tasks assigned, they cared about nothing else". Moreover, activities such as the long-term endless sessions of studying political documents, public accusation meeting for class struggle, and herders' speaking bitterness meeting also affected the timely migration of the pastoralists. It was also pointed out in the official paper that many "livestock which did not move to the summer pastures destroyed the winter pastures and harmed the (rotational grazing) system."[14] In addition, because the pastoralists cared for different herds and flocks in accordance with the arrangements of the production team, they were not familiar with the conditions of individual livestock. The previous intimate association between the herders and livestock disappeared, which prevented the herders to provide targeted cares to the livestock.

In short, the poor body conditions of the undercared livestock increased their risk of dying of snow disasters. Fortunately, the eastern part of the Qinghai-Tibet Plateau had not been hit by any snowstorm for more than a decade since 1954. However, bad luck struck at last. Snow disaster

---

[14] H. Niu. Op.cit., p.283.

in the spring of 1970 resulted in the death of 70 000 livestock in Henan County, and 154 000 in Zeku County. The two counties was hit by another snow disaster in the winter of 1974 and the spring of 1975, which killed a total of 130 000 livestock in Zeku County and 95 000 in Henan County. The herder dpa'-rdo reminisced, "I was caring for a flock of 300 two-year-old lambs that year. It snowed continually from October 15 on the Tibetan calendar to the New Year. Finally, only 30 lambs survived." The new disaster prevention system was conspicuously vulnerable[15].

Two snow disasters exposed the ineffectiveness of the disaster-resistant system. One of the "Five Major Pastures" in China, Qinghai Province was an important livestock producing area, whereby the development and safety (disaster resistance capacity) of livestock husbandry had a bearing on the national supply of livestock products, and the improvement and enhancement of the pastoralist production technology had a bearing on the national economic development. Therefore, the Chinese government attached much attention to the exchanges of disaster prevention and mitigation technology among major livestock producing areas, which expanded the geographical scope for the reference of disaster experience.

As a result, the practice of *Küriyen*, that is, Enclosed-Pastures, was introduced from the pastures in Inner Mongolia. Küriyen is a Mongolian word meaning enclosure walls. What is meant by setting up *Küriyen* is to enclose part of the pasture within the walls (there is a variety of the walls according to the materials used, such as stone walls, clay walls, sod walls, wire fences, and so on, all about 2 meters in height), with the aim to protect natural vegetation growth within the enclosure. Such measure was reportedly effective in disaster prevention on the pastures in Inner Mongolia. According to official media, *Küriyen* had four major strengths: "Firstly, via protected natural vegetation growth and by setting up in all four directions clay walls (or sod walls, or others), the natural pastures can be prevented from livestock trampling and restored faster in vegetation. Secondly, when the grass grows high enough, part of it may be collected and stored as dry green forage, and some seeds may be collected. Thirdly, natural conditions permitting, through irrigation and fertilization, cultivation of some feed plants or quality forage may be conducted to gradually

[15] H.Xiao. Snow Disaster Rethinking, Development. Lhasa: the Tibet People's Press, 1999; The China Society of the Qinghai-Xizang Plateau Research. Resources, Environment, and Development in Qinghai-Xizang Plateau: Proceedings of the First Symposium on the Qinghai-Xizang Plateau. Beijing: Science Press, 1992.

achieve the integration of fodder and forage. Lastly, emaciated livestock can be fed in winter and spring, which helps disaster prevention and livestock protection."[16] It is obvious that the rationale of *Küriyen* technique is still to maximize the forage yield of part of the pasture by preventing the disturbance on vegetation growth from the livestock, and meanwhile plant forge in suitable localities. In 1974, the *Küriyen* technique was applied and promoted in the Zeku County, where the Geri Village is located. A year later, the Henan County, where the Mar Village is located, began to promote the technique by following the exemplar practices of the Zeku County. According to statistics, a total of 60 000 mu of pastures was fenced in the Mar Village, while in the Zeku County, a total of 82 000 hectares of pastures was fenced, each head of the livestock averaging 1.16 mu. According to the herders' assessment, "If properly managed, *Küriyen* should be effective in the protection of pasture, but at that time it was difficult for any measure to be properly implemented, even the best one. Built with turf and sod walls, *Küriyen* are easily subject to erosion of wind and rain, and are not effective in stopping goats, sheep and other livestock, which are good at climbing. Mainly due to poor management and maintenance of the *Küriyen*, the results were not satisfactory in spite of arduous work done and many pastures damaged."

Küriyen construction did not yield the desired results. However, the success of one or two places has offered some enlightenment to the government authorities. First, the enclosed natural growth and cultivation is not only helpful in the restoration of pastures and the increase of forage yield, but also effective in disaster resistance in winter and spring. The enclose pastures later serve as pastures for "disaster resistance and livestock protection". Second, the successful cultivation of introduced perennial forage in *Küriyen* fosters the confidence of the government authorities in improving the pastures. In the western Zeku County, along both banks of the b*a-chu* River, which is not far away from the Geri Village, more areas are exploited for forage planting, which are the base of vegetation cultivation in that county.

[16] News Bureau in Huangnan Prefecture, New Achievements of "Enclosed-Pasture" in Zeku County. *Qinghai Daily*, Oct.1, 1974.

4. Household Responsibility System Era: Integration of Property Rights and Government Relief

Household Responsibility System reform has been implemented in rural and pastoral areas since 1979. As the underdeveloped inland areas, the Mar and Geri villages did not start the reform until 1984. The production and management systems have undergone further great changes entailed by the reform: the livestock were to be individually owned by the household; although collective ownership of the pasture is guaranteed by the law, the pastoralists, in some sense, were to have the private property right over their contracted pastures as the long-term contract policy allowed them not only the rights to exploit but also the rights to transfer their pastures contracted. Another notable change was that the divergence of the Mar and Geri villages in their management modes of pastoralism: the pastoralists of the Mar Village adopted the sedentary way of living in 1976 and completely terminated the nomadic life of long-distance migration. Thereafter, the herders practice seasonal rotational grazing only on the pastures contracted to them in 1984 according to the number of family members and livestock. Although the ancient mode of migratory pastoralism remained in the Geri Village, the pastures were also allocated to the herders' households by virtue of the number of family members. The reform in ownership and differentiation in management modes will inevitably affect the herders' strategies and measures for disaster prevention[17].

In the meanwhile, some other factors exerted increasingly important impacts on the pastoralists' mechanisms for disaster prevention and mitigation. First of all, since the 1980s global warming has brought about clear pasture degradation and on-going deterioration of ecological environment in the eastern part of the Qinghai-Tibet Plateau where the Mar and Geri villages locate[18]. Second, as the living standard has been markedly im-

---

[17] G.Harding. The Tragedy of the Commons. - In: Hardin and J. Banden (eds). Managing the Commons. NY, 1977.

[18] Study of G.. Zhang et al notes that weather station data from 1961 to 1996 show that although the total precipitation in the southern Qinghai Plateau maintains the balance, the precipitation in winter and spring is increased, while the rainfall in summer is reduced, which results in the increased droughts in summer, and increased snowstorm in winter and spring. Temperature information study shows that the temperature rising speed in grass-turning-green period gets slower and slower year by year, and the temperature dropping speed in grass withered period gets faster year by year; therefore, the grass-turning-green speed is delayed and the

proved (especially the improved health care conditions and long-term social stability), the past few decades has seen a rapid increase in the local population, which has almost tripled since the 1950s. There is a substantial increase in the total number of livestock, which has increased by fourfold in the Mar Village, and more than doubled in the Geri Village. The deteriorated pastures are more strained by the twofold population increase. Third, due to the emergence of warm winter, the snowfalls in the winter and spring seasons in the eastern pastoral area of the Qinghai-Tibet Plateau have been significantly increased.[19] These changes expose the pastoralists' to higher risks of snow disaster, and render the traditional systems, techniques and countermeasures completely ineffective. Thus, in order to adapt to new changes, the local government authorities and herders have conducted a range of pilot trials on the establishment of new disaster prevention and mitigation systems, techniques and technologies. Although the relevant experiments are still on-going, after more than two decades' exploration, the basic frame and the line of thought have emerged: First, the local authorities tries to trigger the herders' economic rationality through clarifying the private property rights, involving them in rational and far-sighted exploiting of the pastures; at the same time, they purposely direct the herders to increase investment in pasture construction and improvement with the aim of raising forage yield of the pastures through increased investment. Second, with the initial success of economic reforms, the local governments have gradually accumulated the capacity in snow disaster intervention (including the preventive preparedness and disaster

---

grass withering speed is accelerated, thereby the effective accumulated temperature in grass growth period is reduced, and the effective grass growing season is shorted, which leads to adverse effect on both the grass growing and yield. The observational data from the enclosed pasture during 1988 to 1997 indicate that the height of local grass generally decreased by 30% to 50% in 10 years, and the forage yield of natural pasture dropped substantially. See: G. Zhang Research on Climate Change on Southern Qinghai plateau and Its Impact on Alpine Meadow. *Acta Prataculturae Sinica*, No3, 1999.

[19] The studies on the information from Weather Station of Zeku County during 1958 to 1994 show that there is obvious relativity between the spring snow disaster and the warm winter in the region. The six warm winters were concentrated from 1987 to 1994, which indicates that the climate change increased the possibilities of snow disaster in the region. See: K.Ren. Analysis on the Relationship between Warm Winter and Spring Snow Disaster in Southern Huangnan Prefecture. *Qinghai Meteorology*, No3, 1996.

relief). Hence, governments' assistance has become an important supplement to pastoralists' capacity of addressing snow disasters.

The changes have not taken place overnight. In the early period of reform, although pastures were nominally contracted to individual households, the pastoralists did not formulate a clear concept of private ownership. As there was no strict physically tangible devices of separation along the borderlines between the pastures of different households, the herders did not confine their grazing strictly within their own defined pastures, but rather often accessed others' pastures freely. On the other hand, the privatization of the livestock stimulates herders' initiatives in developing larger herds and flocks and pursuing wider margins of market interests. The pastoralists did not determine their size of herds and flocks in accordance with the carrying capacity of their pastures, but fancied to share the pasture of others instead. Therefore, the livestock population of most households steeply increased. Consequently, there arose a sharp incompatibility between forage supply and livestock population. Pasture degradation caused by overstocking was commonplace, and typical effect of "the tragedy of the commons" unfortunately emerged. Statistics clearly reflects the tendency of rising mortality rate from snow disasters during this period. From 1984 to 1994, the annual average mortality rate of adult livestock in the Zeku County was 18%, with the highest rate reaching 27.18%. Hence, according to statistical records it was a decade with the highest mortality rate[20].

The on-going deterioration of ecology and rise of snow disaster risks have awoken the governments' awareness that without tangible devices of pasture division nominal privatization is not exclusive at all, and the private property rights cannot be guaranteed. As a result, it is not likely that the herders will increase investment to preserve and improve pastures; but conversely, their priority will be to constantly increase their livestock population to seek the maximized short-term benefits, which will aggravate the deterioration of pastures. Therefore, only through physically dividing the individual pastures can the private property rights be effectively defined. Hereby, the governments have tried to "deepen the reform", demanding the herders to enclose their pastures with wire fences through providing subsidies and special loans. Policies have been launched, underlining that the valid period of contract will be extended to 30 to 50 years to ensure the sufficient duration for profit after the investment. After carrying

[20] See: Statistics of Death of Adult Livestock in the Zeku County during 1970 to 1995. *Zeku County Annals*, Op.cit., p.139.

out physical dividing between pastures of different households, the governments have also encouraged the herders to divide their own pastures into smaller seasonal ones with fences to implement rotational grazing and pasture restoration. In addition, by virtue of good pilot results, the local governments have asked every household to set aside a plot of pasture for disaster resistance, viz. the pasture for disaster preparedness and livestock protection, targeting the potential hazards of increasing snowfalls in spring. The average area of this type of pasture for every household in the Mar Village reaches 500 mu, where the pastures are relatively expansive, while that for every household in the Geri Village is only 200 mu. The exploitation of this type of pasture is intervened by the government. A case may be taken to illustrate this. In 2007 it was predicted by weather forecasting that there would be more precipitations (snowfalls) in spring in Henan County. Hence, the government instructed that the reserved pasture ought not to be used before March 15 in order to be better prepared for the possible snow disaster in late spring and early summer. This measure receives well among the herders. Those pastoralists of the Mar Village believe that their reserved pastures may still be grazed even the snowfall is as thick as 15 centimeters, and can well-weather a moderate snow disaster.

It is a compulsory requirement of the government that the herders cultivate forage and reserve a certain amount of forage for climatic events. What is distinct from the past is that to avoid destruction of natural vegetation the government generally prohibits the reclamation of grassland to cultivate forage, but rather demand that when the herders move to (or practice rotational grazing in) the summer pastures (camp sites) they plant barley, oats, and other plants for forage use in the livestock sheds on winter settlements, which may be collected and reserved after autumn. It is demanded in the Henan County where the Mar Village is located that each pastoralist household should cultivate forage on an area of no less than 5 mu, and meanwhile, store concentrate feed of no less than 150 kilos. Another case is from the Geri Village. The household of the pastoralist bsod-nams owns a flock of 40 sheep and a herd of 20. According to the government's stipulation each household should have in store a certain amount of forage and feed, which can last for 20 days. In 2006, this household stored 925 kilos of forage and 100 kilos of feed.

At the same time, with the initial success of economic reforms, the revenues of governments at all levels have steadily increased. Although the development in western inland regions relatively lags behind, those regions are obtaining financial subsidies from the central government and financial assistance from economically developed coastal provinces, which enhances the local governments' capacity of increasing the input in shaping the disaster prevention system. The first among the priorities is the construction of warm-keeping livestock sheds. Steady attention has been paid on the construction of livestock sheds since the People's Commune Era, and the livestock sheds at that time were mainly makeshift ones made of clay and wood. Since the 1980s, the government has supported and extended three types of livestock sheds with better heat insulation effect on the basis of pilot application. The earliest was the type with the plastic plate roof in order to address the shortage of timber on the pastures. Later the plastic plate was replaced by the more solid synthetic materials. The latest type of livestock shed is the steel-framed brick ones, which may serve the double functions of planting vegetables in summer and keeping the livestock warm in winter. It is reported by the herdsmen that within the closed livestock shed, sheep can survive for six days even without any feed, but otherwise, they can only survive three to four days. The government also provides herdsmen with subsidies according to annual proportions. blo-bzang's family from the Mar Village possesses a livestock shed with an area of $200m^2$, for which the family has invested 6700 Yuan (RMB) and the government has allotted 3000 Yuan as subsidies. When the present writer made the investigation in 2007, those herdsman families that had finished the construction of warm-keeping livestock sheds in Henan County approximately accounted for 50%. Besides, the government also increases the financial expenditure proportion to reserve forage for disaster relief, which will be offered to disaster-stricken area. If fortunately there is no harsh climatic event, the reserved forage will be sold to herdsmen at low prices in April.

Some herdsmen from the Geri Village follow an alternative route of development, which is rather representative in the new era. In order to halt the tendency of ecological deterioration of the Qinghai-Tibet Plateau, the Chinese government has put in huge investment to initiate the "Three Riverheads (referring to the Yangtze River, the Yellow River, the Lancang River) Ecological Protection Program". One of the important strategies is to resettle the herdsmen residing in critical ecological region with a view to help the pastures there fully recuperate and restore. Due to its location in the valley of the Geri River, a tributary of the Yellow River, the Geri Village is within the area of ecological protection and resident resettlement. In order to reduce the number of livestock (mainly through imposing grazing ban or advocating rotational restoring), the government resettle half of the pastoralist households (100 household) from the Geri Village to the *ba-chu* River region with relatively better traffic and climatic conditions, which was the most successful area of forage cultivation in the era of the People's Commune. The government provides them with housing, livestock sheds, and a small piece of pasture; besides, every household obtains nearly 1000 kilos of forage from the state-run forage cultivation base every year. Thus, through these measures it is ensured by the government that they can still engage in small-scale livestock husbandry. At the same time, the government encourages them to develop alternative economies. The 41-year-old blo-bzang and his wife exert themselves to adapt to the new life: the husband joins in the construction team to pave the road, while the wife is involved in raising fast-growing sheep at home, planting vegetables and peddling on the side of the highways. Hence, off-livestock earnings predominate in the household income. The senior herdsman sangs-rje' courtyard is near the highway. Taking this advantage, he builds more rooms to rent and brings in considerable income. Although still engaged in raising the livestock, they gradually become agriculturalists, and sheltered rearing liberates them from any concerns about snow disasters. However, they are confronted with new challenges, viz. to extract more profit margins, they have to learn the more precise cost-profit calculation and focus their attention on the market price of the feed, forage and livestock products. Their way of life is fundamentally different from traditional pastoralism. The collective mutual assistance and glory as well as the migration in snow have gradually been reduced to lingering historical memory.

In short, setting up the pasture fences (including disaster resistance and livestock protection pasture), building houses for the herdsmen, constructing livestock shed as well as cultivating and storing forage are summarized as "Constructions of Four Matching Sets", which are the core of snow disaster prevention system advocated by the government in the new era. The new disaster prevention system highlights the importance of the public service of government; the government not only develops relevant rules, but also makes it compulsory that the herdsmen should conduct the self-insurance (such as the establishment of disaster resistance and livestock protection pasture) so as to minimize the tragic disasters resulting from the herdsmen's leaving-things-to-chance mentality. The privatization of pastures and livestock may not necessarily bring about the effect of exploiting the pastures in accordance with the carrying capacity of pastures, however, it stimulates herdsmen's initiative to protect pastures and attend livestock, and the anticipation of utilizing the pastures in a sustainable way also oblige them to control of the size of herd, and even population growth, which undoubtedly plays a vigorous role in enhancing their disaster resilience. However, an inevitable result of privatization is the division of snow disaster prevention capacities among herdsman households; the households with proper management abilities and better economic conditions not only have good self-protection capacities, but also are prioritised to enjoy various preferential policies provided by the government according to the current supporting policies (almost all poverty-alleviation projects require the herdsmen to provide matching funds), whereas on the contrary the pastoralist households with poor management capacity are threatened by higher risks of snow disasters. Moreover, privatization will bring about class distinction, and the pastures will likely fall into the hands of only a few people through the constant market circulation, which will raise challenges to the social ideal of justice pursuing.

From April 29th to May 7th in 2004, the Henan County and Zeku County encountered constant snowfalls, with the average thickness reaching 20 centimeters, and the pastures were underneath thick snow cover for 7 to 10 days on end. As a result, 30 000 livestock were killed in the Henan County, accounting for about 4% of the total livestock population, while in the Zeku County 34 000 heads of the livestock were killed. Although the livestock mortality rates decreased significantly, the losses were still enormous. Dpa'-rdo, the herdsman from the Mar Village, reported that his household had 260 sheep and 100 lambs at that time, and the snow disaster killed 20 adult sheep and 60 lambs. In the same village, the household which suffered most severely lost 70 yaks. The authorities' investigation and study showed that the losses may be attributed to partial implementation of effective prevention measures due to lack of financial capacities of many pastoralist households, while the government's subsidy remained difficult to cover large areas in the short term. Therefore, improving and enhancing the capacity of snow disaster prevention is still the major challenge for the local government and herdsmen.

5. Discussion and Policy Proposals

The distinct natural and geographical conditions of the Qinghai-Tibet Plateau predetermines that the snow disaster risk is the eternal challenge for nomads in their production and life. Therefore, the how-to's of addressing snow disasters are inherent in their social system and production techniques.

The disaster prevention systems and technologies in tribalism era focused on sharing collectively possessed pastures and taking full advantage of natural pastures, which was the reasonable choice made by the local pastoralists in their thousands of years of historical development according to the comprehensive factors involving local social and natural environment as well as lack of outside intervention. Snow disaster prevention systems and technologies were integrated in the nomadic culture which lasted for thousands of years, fossilizing the nomads' understandings of and reflections on the relationship between human beings and nature on the plateau. However, there were obvious limitations inherent in the disaster prevention system prevailing in the tribalism era. When the disasters were too severe to endure, significant economic losses and even personal casualties were inevitable. Although the fluctuations in the losses of the human and livestock population might well result in the balance between man and nature, which was critical for the effectiveness of the traditional systems, the senseless loss of the results of human labor was a tragedy in any way.

The disaster prevention systems in the era of the People's Commune were characterized by conspicuous inserting of external practices, experiences and technologies. Although such prosaic systems paid its due expense, the introduction of new technology enriched the prevention repertoire of herdsmen. However, to some extent the negative effect of the public ownership of the means of production and equalitarianism in income distribution offset the positive effects brought about by new technology.

The market-oriented reform of Household Responsibility System was a passive choice rather than a vigorous innovation. The changes in the ecological environment and rapid expansion of human and livestock population force the herdsmen to conduct the increasingly sophisticated operations of pastures, and even to raise investment to increase the output. Fortunately, with the gradual increase of China's comprehensive national strength, the government takes disaster prevention and mitigation as an important component of the public services, and the increasing national inputs may to certain extent make for the inadequate capacity of herdsmen.

As the government has already determined to build livestock sheds for every household in the future, the biggest problem for snow disasters' prevention at present remains the lack of livestock feed after snow. As mentioned above, the traditional methods based on natural resources have become inadequate due to the degradation of pastures and the increasing number of livestock, and the settlement of herdsmen worsens the conditions of the pastures. The study of relevant government departments indicates that the pasture contract system and settlement[21] renders the "grazing area smaller, and rational rotational grazing and seasonal pasture shift impossible. The long-term repeated grazing and trampling on fixed pastures result in constant intense overgrazing. Thus, the forage is incapable of regenerating and the original vegetation of the pastures is damaged, which results in the rampant growth of poisonous weeds, paving the way for the spread of harmful rodents. The pasture is in severe or extreme degradation."[22] The change in ownership will inevitably trigger comprehensive responses; therefore, the proper way of thinking when addressing any problem should be to take everything into consideration. To the present author, there are only two ways to solve the problem of insufficient feed to enhance the herdsmen's capacity to prevent snow disaster: one is to restore the natural yield capacity of pastures, and the other is to increase the output through the cultivation of grass and herbage.

It is the most economic way to depend on the natural capacity of pastures to prevent snow disaster. There are two options to achieve this goal. On the one hand, the settled herdsmen should be allowed to conduct the nomadic grazing once again. Although rotational grazing is conducted by herdsmen on their own contracted pastures, experiences tell us that divisive pastures are particularly vulnerable to rodent infestation, and the rats accustomed to inhabiting in the low grass area may also intrude the adjacent tall grass area. So only by practicing rotational grazing on large areas of pastures may the desired ecological effects be substantially achieved. The feasible solution is to expand the area of rotational grazing through merging the pastures of different households or even those of the entire village, and to restore the seasonal nomadic migration, but the reasonable factors of the contract system should be inherited (such as being conducive to resolve pasture disputes, and protect the interests of the poor), and the private property rights should be retained as well. The herdsmen may be combined together through pooling of pasture as shares. The carrying capacity of the pasture of each household should be estimated, and then each household may be paid or charged on the basis of the size of their herd and

---

[21] The Government has been encouraging the settlement of herdsmen because the scattered residence causes the high cost in the development of all social undertakings (such as education, health care, and culture), and the development of the local community is hindered to some extent.

[22] H. Ma. Pasture Supervision Station in the Zeku County of Qinghai Province. - In: All the Natural Grassland of the Zeku County Contracted to Herdsman Households. *Qinghai Prataculture*, No2, 2003.

flock. On the other hand, where the pastures deteriorate seriously, grazing is to be banned through the government's purchasing the right of pasture exploitation. The current ecological resettlement is a case in point, and the pastures are to be used again only when restored.

According to the current population, livestock scale, and the carrying capacity of the pasture, the pastoralists' needs of being well prepared for the snow disaster can not be satisfied merely to restore the natural capacity of the pastures, but the herbage yield of pastures need to be increased. There are at least two alternatives. The first is that many years' experiences show that most of the rangelands are not suitable for land reclamation and forage planting, and random reclamation shall result in ecological disaster. However, that does not mean giving up cultivation forever. On the contrary, forage planting technology which does not destroy the original vegetation should be encouraged. In fact, related pilot trials have been conducted for many years, but there are still many technical problems urgently awaiting solving. The direction of future development of pasture husbandry is to foster a strain of herbage that is adaptive to the plateau climate and the traits of plateau vegetation, and explore the cultivation techniques and preservation devices of such grass. Besides, technology-intensive forage industry is to be developed. The next is to gradually settle the herdsmen in the areas suitable for reclamation and cultivation so that they may develop the agriculturalist husbandry or other alternative economies to enhance production efficiency. Nomadic livestock husbandry to be replaced by agriculturalist husbandry is regarded as the necessary consequences because under the premise of the given amount of resources, population growth will inevitably lead to diminishing returns, while the increase of livestock population will result in excessive use of resources. However, the present writer believes that such suitable areas are few in nomadic rangelands. At the same time, it needs to be pointed out that high social costs are inherent in such practice, which may cause drastic cultural changes. Whether the cultural adaptation of herdsmen is successful or not will exert a direct impact on the stable development of the local community.

References

Dong W. Analysis of Climatic Characters of Snow Disasters on Pastures on East Qinghai-Tibet Plateau. *Plateau Meteorology,* No4, 2001.

Ekvall, R. B. Fields on the Hoof: Nexus of Tibetan Nomadic Pastoralism. New York, Holt, Rinehart and Winston, 1968.

Goldstein, M. C. and C. M. Beall. Nomads of Western Tibet: the Survival of a Way of Life. Hong Kong: Odyssey Productions Ltd., 1990.

Gong D., M. Hao, Synthetical Indexes of the Disaster of Snow Cover. *Journal of Applied Meteorological Science*, September, 1998.

Harding, G. The Tragedy of the Commons. - In: Hardin and J. Banden (eds). Managing the Commons. New York: W. H. Freeman and Company, 1977.

Henan County Annals, by the Local Records Compilation Committee of Henan Mongolian Autonomous County. Lanzhou, 1996.

Local Records Compilation Committee of Henan Mongolian Autonomous County. *Henan County Annals*. Lanzhou: Gansu People's Publishing House, 1996.

Ma H. All the Natural Grassland of the Zeku County Contracted to Herdsman Households. *Qinghai Prataculture*, No.2, 2003.

Niu H. Historical Data on Cooperative Economy in Qinghai Agriculturalist and Pastoralist Areas. Xining: Qinghai People's Press, 1993.

Ren K. Analysis on the Relationship between Warm Winter and Spring Snow Disaster in Southern Huangnan Prefecture. *Qinghai Meteorology*, No3, 1996.

Qin N. Rebuilding Historical Climate Data and Research on Climate Changes in Qinghai Province. Beijing: Meteorological Press, 2006.

Qinghai Daily, 1949-2007.

Shi G. Natural Disasters in Qinghai. Xining: Qinghai People's Press, 2003.

Sun W. & Wu Y. Snow Disaster Sequence Chronology and Climatic Statistical Characteristics of Southern Qinghai Plateau. *Qinghai Meteorology,* No.1, 1996.

The China Society of the Qinghai-Xizang Plateau Research. Resources, Environment, and Development in Qinghai-Xizang Plateau: Proceedings of the First Symposium on the Qinghai-Xizang Plateau. Beijing: Science Press, 1992.

Wen K. Chinese Meteorological Disaster Dictionary: Qinghai Volume. Beijing: Meteorological Press, 2007.

Xiao H. Snow Disaster Rethinking, Development. Lhasa: the Tibet People's Press, 1999.

Yu X. Social Investigation on Nomadic Tibet in the Northwest of China. Beijing: The Commercial Press, 1948.

Zeku County Annals. Beijing: China Yearbook of the Counties & Towns Publishing House, 2005.

Zhang G. Research on Climate Change on Southern Qinghai plateau and Its Impact on Alpine Meadow. *Acta Prataculturae Sinica*, No3, 1999.

Zhou L. The Basic Characteristics of Heavy Snowstorm Process and Snow Disaster Distribution in Eastern Pastoral Areas of Qinghai-Tibet Plateau. *Plateau Meteorology*, No4, 2000.

Illustrations

1. Female Herders in Henan County

2. Glimpse of an enclose pasture in Henan County for disaster prevention and livestock protection

3. Interior of a new style livestock shed in the Zeku county

Abstract

Observations conducted on the rangelands on the Eastern Qinghai-Tibet Plateau (the Henan County and the Zeku County in the southeast Qinghai Province) indicate that the following conditions result in sure snow disasters together with heavy snowfall. Firstly, the herbage height is undergrown. Besides natural factors, this undergrowth of forage chiefly results from improper exploitation of the pastureland set aside for winter grazing. Consequently, when the withered forage is covered deeply underneath snowfall, the livestock are starved to death because of dearth of feed. Secondly, it is severely cold after snowfalls. Coldness not only freezes weak livestock to death but also starves healthy livestock as it prevents accumulated snow from melting. Thirdly, the government's weak capacities in such efforts as disaster prevention and relief cannot improve the situation of feed scarcity. Therefore, it can be seen that snow disaster risks are directly associated with the exploitation system of pastures and public service level of the government.

The risks are both related to natural and human factors, that is, different ownership systems of pastures and public service levels of the governments, which vary greatly in various historical periods. Rational system and perfect public service are crucial to reduce disaster risks.

Prof. Zha Luo (bkra-shis don-grub), Ph.D., Associate research fellow with the Section of Ethnic Minority History of the Institute of Ethnology and Anthropology under the Chinese Academy of Social Sciences. He is involved in research of Tibetan area development, cataloguing classical Tibetan works in the Library of Institute of Ethnology and Anthropology. Publication: *Under Pipal: Study on Tibetan Buddhism Culture (1997, 2003).*

ST ELIJAH, THE RULER OF SUMMERTIME NATURAL DISASTERS

RACHKO POPOV

PROFESSOR, DSC, DIRECTOR OF THE ETHNOGRAPHIC INSTITUTE AND MUSEUM, BULGARIAN ACADEMY OF SCIENCES

Among summertime natural disasters, traditional Bulgarians have always listed drought, torrential rains, and floods; thunderstorms, devastating hailstorms and strong whirlwinds. For them, these summertime elements were of demoniacal origin and nature; they were brought about by supernatural mischievous mythological personages such as dragons, lamias, elves and vampires. Their main opponent, according to Christian conceptions, was St Elijah. This was so, since according to the myth about dividing the world among the saints - preserved mainly in Bulgarian folk songs - the Lord assigned St Elijah to be the patron of the upper sphere of the tripartite cosmic space; he was the sovereign of heaven. He controls clouds, fogs, winds, and rain; thunders, lightning, and hail. That is why throughout the Bulgarian lands he has been called *gradushkar* (ruler of hailstorms). The folk imagination and folklore describe the saint as an old man with a long white beard:

And when St Elijah heard,
He rose and shook his old body,
His old body, and his white beard,
And then a thunder echoed...[1]

The saint wears a homespun coat: in one of its sleeves he keeps a handful of hail, hiding in the other his spear – the lightning. His head is often covered with 9 or 12 veils. In this way the Lord does not let him look down towards the earth and unleash the elements of heaven. St Elijah rides across heaven in his fiery chariot, driven by two or four fast, either white

[1] T.Koleva. Kalendarni praznitsi I obichai v Haskovsko, [Calendar Holidays and Customs in the Region of Haskovo]. AEIM, No701-II, p.14.

or bright red, horses. Perched on his right shoulder is an angel, who holds the reins of the horses. In summertime the saint is very busy. He is waging continuous furious battles against the demons of the heavenly elements. Holding pride of place among them are the dragons, the whirlwinds, and the lamias, which, according to ancient pagan conceptions, are the masters of the clouds, lightning, rain and fog. The beliefs that have survived along these lines are particularly diverse. Sometimes the clouds are represented as dragons, while the fog is their sister. Since she sinned greatly, her brothers locked her in a cave. When, once in a while, they let her out, she only moves low to the ground, because she was ashamed of her sins (myth originates from the region of Targovishte). In another instance, the fog is a dragon, coming to abduct some earthly maiden (myth originates from the region of Preslav).

Almost all over the Bulgarian lands, summertime thunderstorms are explained as a battle, in which the saint fights against them. This motif is particularly popular in Bulgarian folklore[2]. St Elijah is the ruler of the winds that he keeps locked in his heavenly cave. In the minds of people, summertime strong winds and whirlwinds are also demoniacal in origin. They are brought about by the souls of the "unclean" deceased (people who have drowned, or have been hanged, murdered, or aborted illegitimate babies), by evil spirits, and wood nymphs, by winged dragons, and lamias, by witches and sorceresses. According to a legend originating in the village of Skelitsa, Yambol region, whirlwinds are middle-aged evil women. They are dressed up in old, shabby, and dirty clothes. They carry a stick, with which they strike people on the heads to make them dizzy, and then sicken them.[3] If thunderbolts are heard on St Elijah's feast day, maidens do not drink water from containers. Otherwise, they believe that a dragon, having come in the form of thunder from heaven, will fall in love with them.[4]

The Bulgarian Karakachans consider that with his fiery arrows St Elijah persecutes the evil wood nymphs[5]. In the course of years, the pre-

---

[2] D.Marinov. Izbrani proizvedenia. [Selected Works], Vol.1, Sofia, 1981, p.681-684; I.Georgieva. Bulgarska narodna mitologiya. [Bulgarian Folk Mythology]. Sofia, 1993, p.109-115; M.Benovska-Sabkova. Zmeyat v bulgarskiya folklor. [The Dragon in Bulgarian Folklore.] Sofia, 1995.

[3] A. Bozakova. Etnografsko izsledvane [further: EI] na s.Skelitsa, Yambolsko [Ethnographic Study of the Village of Skelitsa, Yambol. 1947, p.113.

[4] L.Hristov. EI na s. Podgoritsa, Targovishko. [Ethnographic Study of the Village of Podgoritsa, Targovishte Region.]. RKS [A Manuscript of Students' Course Work by the Dept. of Philology, Sofia University], No70, 1937, p.67.

[5] Zh.Pimpireva. Karakachanite v Bulgaria. [Karakachans in Bulgaria.] Sofia, 1998, p.117.

Christian, pagan, demonic personages have very often been integrated into one, forming the group of the so-called *dyavoletini [devilish tricks], pogantsi [heathens], vragove [enemies]* or *bessove [rages],* whose main spokesperson and leader (*tartor*) becomes the Devil (*Satanail, Velsevul, Juda, Anti-Christ*). Christianized legends about the saint began to present him as a tireless persecutor of the Evil One (also referred to as *one-legged, limping, short, horned, tailed, sly*). With his main weapon, the lightning, the saint is always in pursuit of the Lord's antagonist, who is hiding between the horns of animals, or amidst the branches of trees. That is why folk beliefs provide for a ban on shepherds and ploughmen standing close to livestock, or hiding under trees, in the event of summertime thunder storms. According to a legend originating from the region of Byala Slatina, St Elijah was the only saint who ascended to heaven alive. Before that he had preached the Christian faith on Earth. But it happened so that he killed his father, suspecting him of a family treason. He wanted to repent, but only the Devil, dressed up as a monk, agreed to hear his confession. Then the Devil made him light a fire on top of a rock overhanging an abyss, and jump across it, causing his death. But an angel of Heavens flew down; he took Elijah by the hand and led him away. Ever since that time, St Elijah has been constantly persecuting all the devils on earth with his lightning and thunder[6] (Illustration 1). Bulgarians from the region of Uzunkyupriya, in turn, believe that St Elijah holds a lightning bolt in his left hand and thunder and rain in his right hand. When he persecutes the Devil, he first sends a lightning bolt after him and then thunder. If the Devil has hidden amidst the animals or in a tree, St Elijah sets them on fire.[7] In wintertime, St Elijah "puts on a sheepskin coat and stays inside in the warm". He is a fairly chilly old man. Moreover, he is occupied by other affairs. During that time he follows and instructs the souls of the sinful dead in their celestial workshops. Because of their sins on earth, these souls are destined and punished to make, day and night, the snow, the frost and the hail[8]. They fill the saint's bags, barrels and cart with ici-

[6] M.Ivanova. EI na s. Gabare, Beloslatinsko. [Ethnographic Study of the Village of Gabare, Byala Slatina Region.]. RKS, No330, 1938, p.122.

[7] G.Nikolov. EI na s. Chopkyoi, Uzunkyupriisko. [Ethnographic Study of the Village of Chopkyoi, Uzunkyupri Region.]. RKS, No131, 1940, p.40.

[8] There are also other beliefs, according to which hail is made in the netherworld by people to whom no memorial services have been dedicated: G.Ivanov. EI na s. Malo Babino, Vrachansko [Ethnographic Study of the Village of Malo Babino, Vratsa Region]. RKS, No313, 1942, p.95, or by dead women who were married to widowers, because "they had made their husbands' previous wives weep in the netherworld": B.Tsvetkova. EI na s. Vardim, Svishtovsko. [Ethno-

cles, which he is to spread on the fields and gardens of the farmers who do not honour him during the summer. The domination of St Elijah over the summer is also obvious from a folk tale from the region of the town of Veles. It runs that once upon a time people asked the saint to beg for them an eternal summer from the Lord. But in the hot weather a great number of grass snakes and dragons thrived, who wanted to marry earthly maidens. Therefore, the Lord was forced to bring back winter again.[9]

Though not so frequently, some folk tales present St Elijah as a shepherd, who could magically play his mellow wooden pipe. And the people often called the summertime white clouds in the sky "St Elijah's sheep". Once the saint was grazing the flocks of a Turkish landlord. But he fell asleep in the meadow and the sheep scattered. As soon as he awoke, the saint prayed to the Lord to bring back to him the sheep and promised a sacrificial offering, should the Lord fulfill his request. And so it happened. For this reason the housewives make and hand specially kneaded round loaves of bread and rolls to all the shepherds in the village on the feast day of St Elijah. According to another folk tale, the saint wondrously developed the *gardjet* tobacco variety. He had been a close friend of the Turkish pasha. He once invited the pasha to visit him and started showering lightning and thunder to show him his skills. The frightened pasha hid in a barrel and, extremely frightened, "pissed in it". In time, when people washed the barrel and dumped the dirty water from it, the new tobacco variety started to grow on the spot.[10] In another variant, widespread in Macedonia, the antagonist of St Elijah was Mohammed himself. Once, during their first encounter, the two "chiefs" decided to try their forces in a fair duel. But Mohammed lost. When St Elijah flew over the mosque and its minaret tumbled down on the order of his enemy, he was not in the least frightened. But as soon as the saint started thundering and sending lightning from heaven, Mohammed was frightened and "shat in his pants" out of fear. Therefore St Elijah turned his excrement into tobacco, "so that people would drink it everywhere, but in the House of God", no matter whether it was a church or a mosque.[11] Bulgarians celebrate St Elijah's feast day (July 20) as one of the greatest and most solemn summertime holidays. It

graphic Study of the Village of Vardim, Svishtov Region.]. RKS, No171, 1940, p.108.

[9] Sbornik za narodni umotvoreniya i narodopis (further SbNU). [A Collection of Traditional Culture and Folklore.]. Sofia.Vol.9, p.169.

[10] N.Eremieva. EI na s. Byala Rada, Vidinsko. [Ethnographic Study of the Village of Byala Rada, Vidin Region.]. RKS, No140, 1945, p.201-202.

[11] E.Lafazanovski. Makedonskite kosmogoniski legendi. [Macedonian Cosmic Legends.]. Skopje, 2000, p.160.

is often considered to be mid-summer, because the saint "puts on his sheepskin coat and goes to the Lord to pray to him for winter." People living along the Black Sea coast and along the Danube observe a ban on bathing on the feast day of St Elijah, because, according to beliefs, at that time "waters turn round" and then St Elijah takes a sacrifice (someone is drowned). By force of tradition, on St Elijah's feast day, the oldest cock of the preceding year (*peinik [singer], the father of cockerels*) is slaughtered as a sacrificial offering to the ruler from heaven. It has long been known to researchers of mythology that the cock is a zoomorphic sun symbol and his being sacrificed to honour St Elijah is completely natural and justified. In northwest Bulgaria women stick a feather from the cock's tail over their gates, to "ward off evil and magic spells". Almost every village organizes common celebrations and gatherings, a mandatory element of which is the *kourban* sacrificial offering. A big male animal is always slaughtered (an ox, a bull, a wether or a ram), purchased by common contributions. The sacrificial offering is made in some holy place within the lands of the village – under the branches of an age-old oak tree, by the side of a mound, a holy spring, a stone tombstone or a chapel, carrying the name of the saint (Illustration 2). In quite a number of cases, the holy site is locally associated with the legend of the self-willed offering of a stag (also a solar zoomorphic mythological symbol), which sometimes comes out on his own to the hunters to become their victim,[12] and in other cases – with the appearance of the mythical host of the village, called *the Holy Proxy, a Valiant Young Man or the Villager* in the south Bulgarian lands.

It turns out that very often the site of the sacrificial offering, connected with the myth of the self-willed sacrifice of the stag, has names of the kind of "Lord's Name", "Church of the Lord", or "Holy Spirit". These names give grounds to some authors to assume that alongside the blood sacrificial offering made there, reflected in them are ancient pagan conceptions connected with the cult for the Thracian heavenly ruler Perkos and his Slavic counterpart – Perun.[13] The sacrificial offering is always sanctified by a priest, who often also performs a service with the request for rain. I would

---

[12] R.Popov. Kurban Sacrificial Offerings on the Feast Days of the Summertime Saints in the Calendar Tradition of the Bulgarians. – In: Kurban in the Balkans. Belgrade, 2007, p.259-268.

[13] G.Grigorov. Kam vaprosa za trako-slavyanskata kultura. Sacronimat *Bozhe ime,* praznitsite *Bozhi duh – Lelino zdrave* I legendata za samovolnata zhertva na elena [On the Question of the Thracian-Slav Culture. The Holy Name *God's Name,* the *God's Spirit – Aunt's Health* Holidays and the Legend About the Self-Willed Sacrifice of the Stag.] – In: Etnos i mentalnost [Ethos and Mentality.] Sofia, 2004, p.15-25.

like to quote an example: in the village of Dryanovo, Popovo region, St Elijah is the patron saint of the local church. In the past, every year 9 or 11 barren sheep used to be slaughtered in his honor in the churchyard. After the meat of the sacrificial offering had been cooked, the bell rang solemnly to mark the beginning of the festive meal for the entire village. Every woman was sure to bring a boiled chicken and round loaves of bread of *new grain.* The priest used to hold a holy service for the sacrificial offering and for the feast, whereby burning candles were stuck on all the round loaves of bread. Then everyone would pass in front of the icon of St Elijah, which had been taken out of the church and placed on a temporary altar. They would bow to it or kiss it and leave money for the church.[14]

What is noteworthy is that in Central West and North West Bulgaria, St Elijah is foremost in the kind of statistics, connected with the saints who patronise family and village stone ritual crosses and holy sites. The saint and his "brothers" Peter and Paul are often described on the stone inscriptions as *guardians of the field.*[15] In view of the modern presentations of the popular tradition, I would like to add here that the family and settlement sacrificial offerings honoring some saints and their sacrificial crosses have been revived after the political changes of 1990 and continue to be partially observed to this day.[16]

The slaughtering of a sacrificial animal on the feast day of St Elijah is also common for most of Moslem Bulgarian gypsies. They call the holiday *Ali Gyun*, which means "the day of Ali".[17] Noteworthy in this respect is the holy compound including the tomb of Bali/Ali and the Church of St Elijah, located in the Knyazhevo residential district of Sofia. The holy site is visited on August 2 (according to the Julian calendar), every year, by Christians and Muslims (Gypsies and Turks). Ethnological observations have found that both religious groups make a sacrificial offering for good

---

[14] P.Milenov. EI na s. Drinovo, Popovsko. [Ethnographic Study of the Village of Drinovo, Popovo Region.]. RKS, No182, 1936, p.166.

[15] B.Petrunova, V.Grigorov, N.Manolova-Nikolova. Sveti mesta v Godechko, Dragomansko i Transko. [Holy Sites in the Regions of Godech, Dragoman and Tran.]. Sofia, 2001, p.123-125.

[16] S.Grebenarova. Rod roda ne hrani. Tezhko mu, koito go nytama. [The Kin Would Not Feed His Kith. Pity on the One Who Does Not Have It.] – In: Narodna kultura na balkandjiite [Traditional Culture of the Balkan Mountain People.] Vol.1, Gabrovo, 1996, p.150-152; M. Benovska-Sabkova. Obrazite v Severozapadna Bulgaria: vid religiozno povedenie v post-sotsialistichestko vreme. [The Votive Sites in Northwest Bulgaria: A Kind of Religious Behaviour in Post-Socialist Times]. *Bulgarska Etnologia* [Bulgarian Ethnology], No1-2, 1999, p.70-84.

[17] E. Marushiakova., V.Popov. Tsiganite v Bulgaria. [Gypsies in Bulgaria]. Sofia, 1993, p.176.

health.[18] Analogous is the case of the Gypsies, who are Christians and Muslims and who pay mass visits to the Monastery of St Elijah or *Aliguni* near the village of Mirkovtsi in the vicinities of the town of Skopje. They spend the night there; they pray for good health and slaughter cocks or lambs to honor the heavenly thunder-ruler, who "persecutes and kills the sinful with his lightning".[19]

The traditional Bulgarian customs against drought, called *Peperuda [Butterfly]* and *German,* or a church service for rain, are often performed specially on the feast day of St Elijah. And according to a folk tale from the region of Razgrad, "once upon a time, St Elijah and Demir Baba were brothers". The shrine of the Shiite saint was close to the village of Voden and was frequented by Christians and Muslims on the feast day of St Elijah. "In it was the tomb of the saint, his bronze shoes and his huge knife; the gate to the tomb was decorated with multicolored threads and above it – on the rock - the big footprint of the saint could be seen – just as Krali Marko's".[20] The "fraternal" relation between St Elijah and Demir Baba could be assigned to the analogous characteristics and functions that both possessed. In the conceptions of the Alians (Bulgarian Shiites), Demir Baba was a legendary hero of a huge stature and valiant soul. He could cross from one hill to the other in one stride. Sometimes, like "the first people on earth" – *the Hellenics, the Hebrews* or *the dog-headed [monsters],* he stumbled into the thorny bushes of the blackberry, because of his three-meters tall stature. Moreover, all the time he fought battles against the lamias and *Shaitan* (the Devil).[21]

Besides the feast day of St Elijah, sacrificial offerings honouring the heavenly Christian ruler of lightning are also made in critical situations, connected with long-lasting natural disasters like hailstorms, torrential rain or drought. Particularly interesting, in this respect, is a story, recorded in the village of Velichkovo, Pazardjik region, in the mid-20th century. As soon as a hailstorm came, three men - widowers as a rule - had to steal one *entirely-black-skinned lamb* from some sheepfold and three pots from some potter. Then they gave these to the local priest. He slaughtered the

[18] L.Mikov, T.Kmetova. Sofiyanetsat Bali efendi – myusyulmanski svetets of XVI v. [Bali Effendi of Sofia – A 16th Century Muslim Saint.]. *Bulgarska Etnologia* [Bulgarian Ethnology], No2, 1995, p.31-50.

[19] T. Petrovski. Kalendarskite obichai kaj romite vo Skopije I okolinata. [Calendar Customs of the Roma in Skopje and the Region.] Skopje, 1993, p.85, 149.

[20] N.Nenov. Patevoditel iz predaniyata na Rusenskite Lomove. [Guidebook of the Legends of the Rusenski Lom Rivers.] Sofia, 2001, p.29-30.

[21] B.Aleksiev. Folklorni profili na shiitskite svetsi v Bulgaria [Folklore Conception of the Shiite Saints in Bulgaria]. Sofia, 2005.

lamb in the churchyard. He filled the three pots with the blood of the victim. Then four men went to the borderline of the neighboring village "where the others' land started". They selected just such a place, "because the hailstorm always hits next to a border". They buried the pots with the blood in the earth and soon afterwards the hailstorm subsided.[22] According to other ethnographic sources, the villagers took the sacrificial lamb to some high rocky locality, where there were eagles' nests. It is generally thought that *cross* eagles fly across the hailstorm clouds and direct them. Therefore, from a global outlook, the eagle is a kind of an ornithomorphic symbol of the thunder saint himself. Disrespect of St Elijah entails grave after effects. If the villagers fail to make a sacrificial offering to him, if they work on his feast day, then a lethal drought will befall the settlement. Then the young men in West Bulgaria would gather in the night time and, completely naked, would wield sticks in their hands and chase away from the village the noxious dragon, who had caused the drought.

According to Bulgarian folk legends, St Elijah has his "sisters and brothers", who do the work of his assistants in his eternal encounter with the demons of the natural elements. In this respect, the combinations of holy heavenly pairs or triads have been fairly diverse and have an explicit regional or local specificity. I am sure that I would not err if I place first the pair of "St Elijah and his sister", because it is indeed the most widespread. In most of the ethnographic and folklore sources, the role of "the sister" has been assigned to St Marina, who had therefore received the attribute of *Fiery*. The feast day of the saint is the third of the triad of the *Goreshtnitsi* (the three hottest days in summer), and according to the Eastern Orthodox calendar, it is on July 17. The most common fairy tale plot presents the two saints as a brother and a sister. Every year St Marina would quite methodically and rationally refuse to impart to her brother when his (birth) day or feast day is. She knows that should he learn that, he would do unheard of wonders on earth: there would be either a hailstorm, or a flood would befall the people. "Because he is a kind of madcap; he would thunder and he would annihilate the world" the myth originates from the region of Sofia[23]).

Among the population of Macedonia, the female saint has often been known as *Marena.* According to information provided by K. Shapkarev, "St Marena had taken the Devil by the hair". The farmers address her with due awe: "Forgive us, Marena, you kindle fires and you put them out!"

---

[22] E.Peneva. EI na s. Velichkovo, Pazardjishko. [Ethnographic Study of the Village of Velichkovo, Pazardjik Region.]. RKS, No48, 1946, p.202-203.

[23] T.Pavlova. EI na s. Kurilo, Sofiisko. [Ethnographic Study of the Village of Kurilo, Sofia Region.]. RKS, No112, 1948, p.148.

The demonic personage of Marena has been mentioned in a great number of folk maledictions of the kind "May Marena kill (strike, take) you!" "May Marena carry you away!"[24]

In the traditional Bulgarian calendar rites and rituals, St Marena is characterized by her fiery (solar) nature. To ward off fire and conflagrations, nothing is done either at home or in the field on her feast day. On that day two twin brothers usually get *live (new, Divine) fire*, which, according to beliefs, guards the village and its residents from disease and epidemics, from wood nymphs and vampires. The saint is glorified in the singing tradition of the Strandja region as "a bride of the sun". She is interpreted as a patroness of the young betrothed, while the Greek Black Sea fishermen worship her as the savior of drowning men.

In the lands, located mostly north of the Balkan Range Mountains, the function of "a sister" of St Elijah is assigned to other female saints. Featuring among them are the Holy Virgin, St Helena, St Mary Magdalena or *Magda*, and St Anne of Summertime. They often carry the definitions *Fiery* or *Burnt.* In the Central Rhodope Mountains in the region of Assenovgrad, these same attributes are given to St Kyriaki – *Fiery* or *Burnt Kyriaki,* whose calendar feast day is on July 7. There are also archive sources, according to which *Quarrelsome Helena, Fiery Marina* or *Burnt Anna* completely substitutes St Marina on her feast day in the calendar, occupying her place of a patron saint of the third hot day.

Obviously, the female personage in the holy heavenly pair is quite movable and is affected by the specificities of the local cult tradition for the saints. In the local variants of the fairy tale plot, commented on above, precisely these saints take the role of "sisters", who do not want to tell their brother Elijah, when his feast day in the calendar is. And, quite naturally, they appropriate some of his solar characteristics, as a result of which they begin to be honored as rulers of hailstorms and the heavenly fire. In the south Bulgarian lands the folk cult for St Helena has analogous features and the saint is universally accepted as a leading "hailstorm saint". In the region of Plovdiv, in Sakar and in the Rhodope mountains it is believed that she, together with her "sister" *Kostadinka* (a female mythological hypostasis of St Constantine) carries the icicles of the hailstorm in the sleeve of her chemise, in her skirts, in her apron or in a basket.

Besides sisters, the chief Christian *thunderstorm saint* Elijah has "brothers". In a folk tale from Panagyurishte, saints Peter, Paul and Elijah are presented as three "brothers by birth". Peter is rich and for that reason

[24] K.Shapkarev. Sbornik ot bulgarski narodni umotvoreniya. [A Collection of Bulgarian Folklore.], Vol.1, Sofia, 1968, p.571.

people celebrate his name day. While the other two are poor, moreover, Elijah is very absent-minded and can never remember when his birthday is: it being during the most strenuous harvest time. Once some peasants met St Elijah on his chariot, while going back from the field. They told him about his feast day, which had already passed. The enraged saint climbed on top of the mountain and in the course of three years did not let a single drop of rain fall on the villagers. Finally, the Lord asked him to have mercy on them and to give them rain.[25] In this case, the fraternal relation of the three saints does not look accidental, in the least, because the twin brothers Peter and Paul, according to folk beliefs, also carry by and large the mythological characteristics of the heavenly fire. St Peter is the master of the Garden of Paradise; he keeps the keys to the gates to Paradise. Together with his brother Paul he brings about summertime fires and hailstorms, especially, if he has not been duly honored by the farmers. On the feast day of St Peter (June 29), a young cock – *a Peter's feast day cock* - is traditionally slaughtered in every house. The farmers make up with it for St Peter's fortnight of fasting.

In other variants of the plot, "the youngest brother" of St Elijah is St Panteleimon (July 27). He has often been referred to as *Water Pantelei*, because he is considered to be the cause of disastrous summertime torrential rains and floods.

Other saints of the Bulgarian traditional calendar have also been assigned heavenly functions of a meteorological nature and, for that matter, they also belong to the hailstorm retinue of St Elijah. These are St German (May 12) and St Bartholomew (June 11). Ancient pre-Christian conceptions, connected with demonic heavenly personages, dominating the atmospheric process, have obviously been preserved in the folk cult of St German. In the Western Bulgarian lands and in the contact zones with the Serbs, predominantly, this pagan personage by the name of *German* is by obligation invited by the hosts to the Christmas Eve table, to be fed and placated. Widespread in the same region is also the conception that the hailstorm clouds are led by three supernatural beings: *deaf German, blind German and dumb German.* As soon as such a cloud appears in the sky in summertime, the women light a Christmas Eve candle outside and drive away the element applying oral magic: "German, go to the deserted forest, where no axe fells trees, no cock sings and no lambs bleat!" And men burn aromatic herbs to "chase away the three Germans to another place".[26] In

[25] S.Bistrekova. Etnografsko izsledvane (EI) na grad Panagyurishte. [Ethnographic Study of the Town of Panagyurishte]. RKS, No87, 1945, p.172-174.

[26] S.Nenov. EI na s. Kitino, Omurtagsko. [Ethnographic Study of the Village of Kitino, Omurtag Region]. RKS, No286, 1936, p.101-102.

the region of Kyustendil, the cloud is sent off by the words: "Come back little cloud *Kuzman*, come back, *German* ..."[27] A folk tale from the region of Graovo runs how simpleton German had been declared a saint after outwitting, with the help of St Elijah, the Lord Himself and St Peter in letting loose the heavenly lightning. In the region of Niš, people believe that *Germa* is the first assistant of St Elijah, without, however, having been declared a saint. As soon as a hailstorm cloud appears in the sky, farmers address a prayer to him in the words: "We beseech you, Germo, if you have decided to let loose a hailstorm, let it loose beyond our border!"[28]

In the western Bulgarian lands, in the region of Plovdiv and at some places in North Bulgaria, St Bartholomew is also characterized by his heavenly and solar nature. According to the local conceptions, on his feast day, "with the help of 12 angels, he turns the solar disc towards summertime" or "he puts on a sheepskin coat and goes to the Lord to bring the snow". On his *weighty feast day* no work is done, in order to ward off hailstorms and summertime whirlwinds. According to another belief, *"Vartolum* turns the sky and the sun" and thereby causes whirlwinds and storms. For that matter, on his feast day men do not rein in the oxen, while women do not plant onions so that "their heads may not spin".[29] At the same time St Bartholomew, likewise, often makes up holy calendar pairs or triads with other saints. The most popular are the following combinations: Bartholomew and German; Bartholomew and Varnava; Bartholomew and Elysee; Bartholomew, Elysee and Vida. In all similar cases, the saints have been believed by the people to be brothers and sisters. They are predominantly worshipped as hailstorm saints, and for that matter their feast days are also called *Varti-udari [Turn-and-Strike], Varti-puh [Turn Eiderdown]* or *Varti-suchi [Turn and Twist].*[30] According to a folk belief from the region of the town of Tryavna, completely resting on the principles of folk etymology, "German had thundered, Bartholomew had turned the clouds, while Barnabas had turned them back".[31]

Though within a restricted area, including the western Bulgarian lands and Macedonia, *St Marko Gradobia [the Hailstorm]* (April 25) enjoys the

---

[27] M.Arnaudov. Studii varhu bulgarskite obredi I legendi. [Studies of Bulgarian Rituals and Legends.]. Vol.1, Sofia, 1971, p.230.

[28] M.Milichevich. Zhivot srba sel'aka [The Life of the Serbian Peasant.] Srpski etnograficheski zbornik [Serbian Ethnographic Collection], No1, 1894, p.123.

[29] K.Shapkarev. Op.cit., p.570-571.

[30] R.Popov. Svetsi i demoni na Balkanite. [Saints and Demons in theBalkans]. Plovdiv, 2008, p.306-308.

[31] M.Arnaudov. Op.cit., p.231.

fame of a hailstorm saint, too. According to traditional folk beliefs, he carries icicles in a basket. No work should be done in the gardens and vineyards, and one should not fall asleep on his feast day. Men do not harness the animals so that no hailstorm might fall in the field.[32] The hailstormy nature in the conceptualisation of St Marko is also typical of Bulgarian Catholics from the region of Plovdiv. On the feast day they organize traditional ritual processions (*alai*) with the intention that "the fields may be blessed" to yield a rich harvest. During the litany, men stick wooden crosses in their fields and attach small crosses made out of wax onto the trees to protect the crops from hail.[33]

* * *

Are these ancient beliefs still alive among the Bulgarians? Press information in 2005-2007, when there were torrential rains in Bulgaria for days on end, resulting in devastating floods, has show that the faith in the mythical beings of yore is still alive. During the horrendous hurricane at the end of May 2007, residents of the village of Krivinya near Sofia told journalists: "This was a whirlwind; it twisted like a dragon from house to house; it thundered mightily and carried away the roofs of the houses"[34]. Old people still call on St Elijah to help them, when it begins to thunder and lightning appears in the sky. They know that then the saint is persecuting the evil dragon with his lightning. But at the same time, other explanations can be heard of the hurricane, connected with the conception of the Lord appearing at the moment of death of every man. Maria Nikolova (aged 59) from the village of Kalekovets, Plovdiv region, who had been swept by the strong wind and in whose courtyard the wind had felled an old pine and a plum tree, stated: "I thought that the Lord had descended on earth to take me away…To me the fallen pine tree was my deceased father, while the broken plum tree was my mother"[35].

Some of the most outstanding values in the life of the traditional Bulgarian are his home and the domestic animals he breeds. That is why, when an old poplar tree had fallen onto the house of Marin Slavkov in the

---

[32] Y.Zahariev. Kyustendilskoto kraishte. [The Kyustendil Region.] - SbNU, Vol.23, 1918, p.157.

[33] A.Yankov. Praznichni shestviya v kalendara na bulgarskite katolitsi ot Plovdivsko. [Festive Processions in the Calendar of the Bulgarian Catholics from the Region of Plovdiv.]. *Bulgarska Etnologia* [Bulgarian Ethnology], No2, 2000, p.62-65.

[34] The *Trud* daily, No140, May 23, 2007.

[35] Ibid.

village of Yoglav, Lovech region, as a result of torrential wind and rain, breaking its roof, the old man refused to go outside. Stuck under the ruins of his home, he had to be forcibly removed by firemen[36]. And in the spring of 2005, to save her small lamb from drowning, Granny Ivanka from the village of Aglen, Lovech region, took the animal in her hands, climbed on top of the table in the middle of the room, which was rapidly filling with water, and raised the lamb high over her head.[37]

In the past as well as today, in extreme situations, help is requested from the local fortune-teller. When in May 2005, the water of the overflowing Skat river took its first victims – four boys - the mother of two of them – twin brothers, turned to a fortune-teller. The fortune-teller spread out the cards to tell the future and predicted that one of the children would survive. Unfortunately, her forecast failed to materialize.[38]

In my view, it is also important, while the problem is being discussed, to point out the effects of the flooding of a number of archaeological landmarks and Christian monuments – churches and chapels. The Cherven Archaeological Reserve near Ruse is cut off from the rest of the world. Near Ruse, again, rains have dug riverbeds 10 m deep into the rock churches of the village of Ivanovo. The Church of St Nikolai (Nicolas) in the village of Dolna Kremena, Vratsa region, has been completely destroyed by torrential rain. Old church mural paintings from the 14th century have been irreversibly lost. The roof of the church in the village of Sadovets, Lovech region, has collapsed. The church bell has cracked in two. In similar situations of natural disasters the clergy also play a part. Priests perform church services and liturgies for healing; they transfer wonder-making icons to places that have suffered natural disasters. Here are some examples. Father Lyuben from the Monastery of St Mina in Sofia performed daily church services, during which he appealed to the congregation to pray for the rains to stop. He took part in every initiative for helping the destitute[39]. With the same protective goal in view, the wonder-working icon of the Holy Virgin Osenovitsa from the Rila Monastery “set out” via Targovishte and Shumen for Varna. On the way public prayers were held all the time to halt the devastating fires in the Sakar Mountain during the summer of 2007. Processions with the wonder-working icon of the Mother of God from the Bachkovo Monastery were organized with the same purpose in mind[40].

[36] Ibid.
[37] The *Trud* daily, No141, May 28, 2005.
[38] The *Trud* daily, No143, May 30, 2005.
[39] The *24 Chasa* daily, No152, June 6, 2005.
[40] The *Trud* daily, No203, July 26, 2007.

Children's answers in the "Children's Reportage" broadcast every evening on the Bulgarian TV were close to the traditional Bulgarian understanding regarding the causes of tempests. On April 23, 2008 children were asked: "Who makes the storm?" Some of them answered in the following manner. "Nature and the air make it." "There is a tempest when two clouds clash" (according to folk beliefs, the two dragons-clouds fight). "There is a storm when Nature is angry with us" (disasters are God's punishment for the sins committed by humans). The last answer had the most modern ring: "The storm is made by the Devil of Tasmania".

This has been just an ethnologist's cursory look at the present-day symbiosis between old pagan and Christian beliefs, and practices of today in coping with extremely traumatic natural disasters like floods, hurricanes and fires.

References

Aleksiev, B. Folklorni profili na shiitskite svetsi v Bulgaria [Folklore Conceptualisation of the Shiite Saints in Bulgaria]. Sofia, Bulgarian Academy of Sciences, 2005.

Arnaudov, M. Studii varhu bulgarskite obredi I legendi. [Studies of Bulgarian Rituals and Legends.]. Vol. 1, Sofia, Publ. by BAS, 1971.

Benovska-Sabkova M. Zmeyat v bulgarskiya folklor. [The Dragon in Bulgarian Folklore.]. Sofia, BAS (second phototype edition), 1995.

—. Obrotsite v Severozapadna Bulgaria: vid religiozno povedenie v post-sotsialistichestko vreme. [The Votive Sites in Northwest Bulgaria: A Kind of Religious Behaviour in Post-Socialist Times]. *Bulgarska Etnologia* [Bulgarian Ethnology], 1999, No1-2, p.70-84.

Bistrekova, S. Etnografsko izsledvane (EI) na grad Panagyurishte. [Ethnographic Study of the Town of Panagyurishte. RKS, No87, 1945.

Bozakova, A. EI na s.Skelitsa, Yambolsko [Ethnographic Study of the Village of Skelitsa, Yambol Region]. Architect Ts. Romanska, 1947.

Eremieva, N. EI na s. Byala Rada, Vidinsko. [Ethnographic Study of the Village of Byala Rada, Vidin Region.]. RKS, No140, 1945.

Georgieva, I. Bulgarska narodna mitologiya. [Bulgarian Folk Mythology]. (Second revised and expanded edition.) Sofia, Nauka i Izkustvo, 1993.

Grebenarova, S. Rod roda ne hrani. Tezhko mu, koito go nytama. [Family Members do not Feed their Own but Pity he who Hath no Family.] – In: Narodna kultura na balkandjiite [Traditional Culture of the Balkan Mountain People.]. Vol.1, Gabrovo, 1996, p.150-152.

Grigorov, G. Kam vaprosa za trako-slavyanskata kultura. Sacronimat *Bozhe ime,* praznitsite *Bozhi duh – Lelino zdrave* I legendata za samovolnata zhertva na elena [On the Question of the Thracian-Slav Culture. The The sacronym *God's Name, God's Spirit – Aunt's Health* Holidays and the Legend of the Self-Willed Sacrifice of the Stag.] – In: Etnos i mentalnost [Ethos and Mentality.]. Sofia, 2004, p.15-25.

Hristov, L. EI na s. Podgoritsa, Targovishko. [Ethnographic Study of the Village of Podgoritsa, Targovishte Region.]. RKS, No70, 1937.

Ivanov, G. EI na s. Malo Babino, Vrachansko [Ethnographic Study of the Village of Malo Babino, Vratsa Region]. RKS, No313, 1942.

Ivanova, M. EI na s. Gabare, Beloslatinsko. [Ethnographic Study of the Village of Gabare, Byala Slatina Region.]. RKS, No330, 1938.

Koleva, T. Kalendarni praznitsi I obichai v Haskovsko, [Calendar Holidays and Customs in the Region of Haskovo]. Archive of the Ethnographic Institute and Museum, No701-II.

Lafazanovski, E. Makedonskite kosmogoniski legendi. Skopje. Kultura. [Macedonian Cosmic Legends. Skopje. Culture.], 2000.

Martinov, A. Narodopisni materiali ot Graovo. [Folklore Materials from Graovo.] – SbNU, No49, 1958.

Marinov, D. Izbrani proizvedenia. [Selected Works], Vol.1, Sofia, Nauka i Izkustvo, 1981.

Marushiakova, E., V.Popov. Tsiganite v Bulgaria. [Gypsies in Bulgaria]. Sofia, Club 90, 1993.

Mikov, L., T.Kmetova. Sofiyanetsat Bali efendi – myusyulmanski svetets of XVI v. [Bali Effendi of Sofia– a 16th Century Muslim Saint.]. *Bulgarska Etnologia* [Bulgarian Ethnology], No2, 1995, p.31-50.

Milenov, P. EI na s. Drinovo, Popovsko. [Ethnographic Study of the Village of Drinovo, Popovo Region.]. RKS, No182, 1936.

Milichevich, M. Zhivot srba sel'aka [Life of the Serbian Peasant.] Srpski etnograficheski zbornik [Serbian Ethnographic Collection], No1, 1894.

Nenov, S. EI na s. Kitino, Omurtagsko. [Ethnographic Study of the Village of Kitino, Omurtag Region.]. RKS, No286, 1936.

Nenov, N. Patevoditel iz predaniyata na Rusenskite Lomove. [Guidebook to the Legends of the Rusenski Lom Rivers.], Sofia, "Rod", 2001.

Nikolov, G. EI na s. Chopkyoi, Uzunkyupriisko. [Ethnographic Study of the Village of Chopkyoi, Uzunkyupri Region.]. RKS, No131, 1940.

Pavlova, T. EI na s. Kurilo, Sofiisko. [Ethnographic Study of the Village of Kurilo, Sofia Region.]. RKS, No112, 1948.

Peneva, E. EI na s. Velichkovo, Pazardjishko. [Ethnographic Study of the Village of Velichkovo, Pazardjik Region.]. RKS, No48, 1946.

Petrovski, T. Kalendarskite obichai kaj romite vo Skopije I okolinata. [Calendar Customs of the Roma in Skopje and the Region.] Skopje, Feniks, 1993.

Petrunova, B., V.Grigorov, N.Manolova-Nikolova. Sveti mesta v Godechko, Dragomansko i Transko. [Holy Sites in the Regions of Godech, Dragoman and Tran.]. IMIR, Sofia, 2001.

Pimpireva, Zh. Karakachanite v Bulgaria. [Karakachans in Bulgaria.]. IMIR, Sofia, 1998.

Popov, R. Svetsi i demoni na Balkanite. [Saints and Demons in the Balkans.]. Lettera, Plovdiv, 2008.

—. Kurban Sacrificial Offerings on the Feast Days of the Summertime Saints in the Calendar Tradition of the Bulgarians. – In: Kurban in the Balkans. Belgrade, 2007, p.259-268.

The *Trud* daily, 2005; 2007.

The *24 Chasa* daily 2005.

Tsvetkova, B. EI na s. Vardim, Svoshtovsko. [Ethnographic Study of the Village of Vardim, Svishtov Region.]. RKS, No171, 1940.

Shapkarev, K. Sbornik ot bulgarski narodni umotvoreniya. [A Collection of Bulgarian Folklore.]. Vol.1, Sofia, 1968.

Yankov, A. Praznichni shestviya v kalendara na bulgarskite katolitsi ot Plovdivsko. [Festive Processions in the Calendar of the Bulgarian Catholics from the Region of Plovdiv.]. *Bulgarska Etnologia* [Bulgarian Ethnology], No2, 2000, p.62-67.

Zahariev, Y. Kyustendilskoto kraishte. [The Kyustendil Region.] – SbNU, Vol.23, 1918.

Illustrations

1. St.Elijah – the Ruler of the Thunders, Lightning and Hail

2. Family and kin- gathering on the Day of St.Elijah under the totemic pine tree – 1934

Abstract

For traditional Bulgarians all summertime natural disasters, such as drought, torrential rain and floods, lightning and thunderstorms, devastating hailstorms and strong whirlwinds, have a demonic origin and character, having been allegedly brought about by supernatural evil mythical personages like dragons, lamias, wood nymphs and vampires. Their main opponent, according to the Christian conception, is St Elijah. This is so, because according to the myth of the distribution of the world among the saints, mostly preserved in Bulgarian folk songs, the Lord assigned St Elijah to be the ruler of the upper sphere of the tripartite cosmic space; he is the ruler of heaven, therefore he controls clouds, fog, winds and rain, thunder, lightning and hail. The article presents detailed research into the celebration of this saint, marked by a festive church liturgy, and with collective sacrifices (mostly of bulls) on high mountain peaks, under century-old trees, or by forest springs, involving everybody in the village: family and friends.

Prof. Rachko Popov has been director of the Ethnographic Institute and Museum at Bulgarian Academy of Sciences for more than 15 years. He is the author of several monographs: *Butterfly and German* (1989) and works in English, *Twin Saints in the Bulgarian Folk Calendar* (1991); *Concise Folk Festal Calendar* (1993); *Bulgarian Folk Calendar* (1997); *The Balkans Saints and Demons* (2002). He is co-author of the encyclopaedic dictionaries: *Bulgarian Mythology* (1994) and *Bulgarian Folk Medicine* (1999). Rachko Popov is editor-in-chief of the series: *Ethnographic Problems of Traditional Culture* and *Ethnographic Studies of Bulgaria*.

EPIDEMICS, SOCIETY AND COUNTRY: A CASE STUDY OF THE HUANGNAN PASTORAL AREA OF TIBETAN AUTONOMOUS PREFECTURE, QINGHAI PROVINCE

FANG SUMEI

PROFESSOR, INSTITUTE OF ETHNOLOGY AND ANTHROPOLOGY, CHINESE ACADEMY OF SOCIAL SCIENCES

Epidemics are different from other diseases in that they are closely related to people's livelihood once they take place, and are likely to bring about a variety of political, economic and cultural problems. Therefore, they have drawn widespread attention from all societies and countries.

The Huangnan Pastoral Area of Tibetan Autonomous Prefecture is composed of Zeku County and Henan Mongolian Autonomous County, chiefly with Tibetan and Mongolian residents. According to our field study, some areas here turn out to be the plague focus, and epidemics such as plagues once broke out. The occurrence of epidemics has much to do with the local physical geographical conditions, ecological environment and social culture. The social system of Huangnan Pastoral Area has changed dramatically since the founding of the People's Republic of China, and the central-local relationship has gradually been integrated with that of the inland. In a similar vein, China is stepping up its efforts in the medical work of Huangnan Pastoral Area. With the improvement in the medical system and public health conditions, and with the establishment and development of the Prevention and Treatment System of Public Diseases, the epidemic situation in Huangnan Pastoral Area has seen significant progress. The outbreak and spread of severe epidemics has largely been brought under control. This article examines the social changes in the pastoral area of minority groups and the central-local relationship in China through a field study of the Huangnan Pastoral Area, Tibetan Autonomous Prefecture, Qinghai Province.

I. Epidemics and Their Characteristics in Huangnan Pastoral Area

Historically, the epidemics that spread in the Huangnan Pastoral Area mainly include: plague, venereal disease, brucellosis, smallpox, typhoid fever, leprosy and a number of others.

A deadly-infectious zoonotic disease caused by the Bacillus yersini pestis, plague is listed as the No.1 Infectious Disease of Class A in the "Law of the People's Republic of China on the Prevention and Treatment of Infectious Diseases." Qinghai is one of the provinces that have suffered the most severe plagues in the country, as evidenced by the once-popular saying that "China will be free from plagues if Qinghai is safe from it" before the 1980's[1]. The major host animal of Bacillus yersini pestis is the Himalayan marmot. In addition, 18 other animals (such as Tibetan sheep, argali, field mice, foxes, dogs, cats and Felis lynx Linnaeus) and 12 kinds of insects can be naturally infected and spread plagues. It is estimated that the epidemic focus in Qinghai province covers an area of about 200 000 square kilometers, with over 600 epidemic points scattered across 33 counties (cities) and 120 small towns. The outbreak of human plagues was reported almost every year in 1958-2001, except in 1972, 1984, 1999 and 2000[2]. While at the same time, newly infected animals and epidemic foci are continuously confirmed and animal plagues occur every year, which has greatly increased the risk of human infection and its long-distance spreading. Related statistics demonstrate that in 1949-1992 the number of pneumonic plague patients in Qinghai accounted for 57.37% of the total number of China, which was much higher than that in other provinces[3].

An investigation conducted by the Chinese government in Zeku County reveals that there occurred 5 plagues in this area in 1908-1955, causing more than 100 people to die. In 1955-1978, no human plagues were reported in Zeku County. In 1978-1987, a total of 12 epidemic foci were identified in the whole county, and a suspected case of plague was

[1] B. J. Zhou. Review on the fifty years for the prevention and treatment of Saint Sebastian's disease in Huangnan Tibetan Autonomous Prefecture, Qinghai Province. *Bulletin of Endemic Diseases*, No3, 2006.

[2] S. D. W. Luo, C. Li, Z. X. Wang. Analyses for the first case of Saint Sebastian's disease in human habitation in Qinghai highland, *Chinese Journal of Endemiology*, No3, 2002.

[3] J. Q. Zhu, etc. Investigations on the Characteristics of the Spread of Saint Sebastian's Disease In Human Habitation and the Control Measurements in Qinghai Province, *Bulletin of Endemic Disease*, No1, Vol.8, 1993.

discovered in the Jiacang Group of Duohemao Commune. In August 1980, a human bubonic plague took place in a winter range in Duohemao Commune, and 2 people were infected. And in 1991, a mouse plague took place in a pasture of Wangjia County[4]. It is documented in the "Lahai Scripture" that in the 19th year of Emperor Qianlong's reign of the Qing Dynasty (1754), a plague broke out in Neiwata area of Henan County as a result of marmot eating by local herdsmen, killing 10 people. A lama from Laboleng Temple of Xiahe County (Gansu province) who came to chant scriptures in the prince's residence was infected, and caused over 100 people to die when he returned to his temple. This was the earliest description of the spread of human plagues in the epidemic focus of marmots in China. After the 1950's, 9 plagues took place in this county in 2 decades thereafter. And in the late 1970's, the local government carried out a number of investigations on natural epidemic foci, epidemiology, gnawers and insects in the county, and identified an area of about 7000 square kilometers with marmot distribution as the natural epidemic focus, 2 small towns as key areas for epidemic prevention and control, and 2 other small towns as key areas for epidemic prevention only[5].

Venereal disease was also widespread in Huangnan Pastoral Area in history, ranging from syphilis to gonorrhea and chancroidal ulcer. These diseases are mainly transmitted via sexual behavior. As venereal disease is highly incidental and hardly curable, it has negatively affected the herdsmen' health and multiplication. It is reported that before the 1950's the general incidence rate of venereal disease in Henen County was about 65.50%[6]. In the 1970's~1980's, venereal disease was largely wiped out in this area. But from the 1990's, venereal disease began to creep back. An epidemiological investigation carried out in Guashize and Duowa towns of Tongren County and Kesheng and Duosong towns of Henan County in July-August 2003 identified 107 venereal disease cases out of 408 subjects (207 men and 201 women), with a prevalence of 26.23%. The disease types included: 68 cases of gonorrhea (63.55%), 17 cases of syphilis (15.89%), 10 cases of nongonococcal urethritis (9.35%), 7 cases of cauliflower excrescence (6.54%), 4 cases of genital herpes (3.74%), as

---

4 *Records of Zeku County* compiled by the Compiling Committee of Records of Zeku County. Beijing, 2005, p.462-463.

5 *Records of Henan County* (the second part) compiled by the Compiling Committee of records for Henan autonomic Mongolia County. Lanzhou, 1996, p.817-818.

6 Ibid., p.817.

well as 8 cases of combined gonorrhea and syphilis. The patients of venereal disease aged from 16 to 62, with a prevalence of 38.41% for Age Group 30-39, 33.33% for Age Group 16-19, and 32.18% for Age Group 20-29. Moreover, there were 60 cases of male patients (50.07%) and 47 cases of female patients (43.93%). The prevalence of venereal disease in this investigation was significantly higher than that (13.06%) in the 1991 investigation in Guashize Town of Tongren County and that (10. 47%) in the 1992 investigation in Duosong Town of Henan County, increasing by 100.84% and 150.53% respectively. This indicates that sexually transmitted diseases show a spreading tendency in remote pastoral areas[7]. Another investigation on venereal disease in Zeku County conducted in the same year reveals that 6 out of 24 people from the high-risk service industry were found positive in USR (syphilis) detection; and 7 out of 20 people were found positive in gonorrhea examination, both with a prevalence of over 20%.

In addition to the local medical and health conditions, the epidemics in Huangnan Pastoral Area have the following characteristics:

1. The Existence of Natural Epidemic Foci and the Influence of Folk Traditions are the Main Factors for the Occurrence of Some Epidemics.

Firstly, as a host animal with higher economic values, marmot has been hunt by herdsmen. In the epidemic foci of Zeku and Henan counties, herdsmen used to hunt marmots and eat their meat, which increases the risk of being infected and plague spreading. As of a rule, the first cases of human plague almost always derived from eating sick or dead marmot. Thanks to the government's awareness-raising activities and stringent measures, a majority of the herdsmen have abandoned the practice of marmot hunting. However, illegal marmot hunting is still going on in some areas.

Secondly, the occurrence of brucellosis diseases is closely related to the living habits of herdsmen. Local residents are fond of eating half-cooked beef or mutton and drinking raw milk. Most of the food contains Brucella, and may cause infection.

Finally, the widespread venereal disease is closely related to the traditional ideas and educational level. Due partly to the nomadic life-style in pastures where the living and health conditions are relatively poor, and

[7] C. N. Li. Epidemiological Investigations on the Venereal Disease among Tibetan in Remote Pastoral Areas in Qinghai Province, *Chinese Epidemiology Journal,* No9, 2004.

partly to the freer gender relations, venereal infection always occurs in herdsmen, causing the spread of sexually-transmitted venereal disease. In an epidemiological study of venereal disease on Tibetans herdsmen that was carried out in remote pastoral areas of Tongren and Henen in 2003, there were 71 illiterates (66.36%), 17 primary schoolers (15.89%), 15 junior secondary schoolers (14.02%) and 2 senior secondary schoolers (1.87%) in the patients. Besides, 36 cases were unmarried and 71 cases were married[8]. These results indicate that the occurrence of venereal disease has as much to do with the traditional ideas as with people's educational level.

2. Social and Economic Changes and the Increasing Exchanges are Important Factors for the Frequent Outbreak of Epidemics.

Generally, Qinghai province has a small number of pastoral areas and a small population. With the economic development and accelerating urbanization in recent years, particularly since the launching of "Improvement of Villages as a Whole" project, many herdsmen in Henan and Zeku counties begin to settle down in neighboring towns or new villages. This is especially so with Henan County, which is characterized by faster economic development and higher population mobility. In particular, some epidemics that used to be controlled, such as venereal disease, hepatitis and pulmonary tuberculosis, begin to prevail again. Studies show that relevant investigation shows that venereal disease in Huangnan Pastoral Area mainly takes place in areas with convenient transportation, since the residents over there are relatively concentrated and large numbers of people keep flowing in, which has inevitably led to an increase in the exchanges between migrants and local residents, and accordingly, an increase in the chances of getting affected. Furthermore, the epidemic foci of marmot on Qinghai-Tibet Plateau used to be kept in isolation for a long term because of its sparse population and obstructed transportation. And yet, these epidemic foci are gradually transforming into an open type, as a result of economic development, transportation improvement and population mobility. Many poachers from Sichuan, Gansu and other provinces enter Qinghai to hunt marmots, and some restaurant owners also engage in marmot dealings by smuggling them from Qinghai and Gansu to Guangdong. Consequently, more and more

[8] H.Z.Wang. Disaster, Society and Modernization: Review Focusing on North Jiangsu during the Period of the Republic of China. Beijing, 2005.

people are involved in the epidemic foci, which has raised the chances of contacting with plague hosts and the breakout of human plagues[9].

II. Prevention, Control and Understanding of Epidemics in the Traditional Society of Huangnan Pastoral Area

In history, the Tibetans and Mongolians in Huangnan Pastoral Area adopted a distinct socio-political system and a nomadic mode of production and life. The central-local relationship was relatively loose, and the medical and health work here remained under-developed for a long time. Until the period of the Republic of China, no medical and health organizations had been set up in Huangnan Pastoral Area by the government, and modern medical service was simply inaccessible to common people. It is documented that Zeku County did not adopt a new remedy for smallpox until 1944. Nevertheless, people needed to give the doctor a ram if a boy was inoculated and an ewe if a girl was inoculated. Since herdsmen could not afford it, few people were actually inoculated[10]. Therefore, local residents mainly relied on social forces in the case of epidemics, including folk Tibetan doctors, temples as well as norms of behavior in their daily life. There were few interventions from central or local governments. Even after the founding of the People's Republic of China, social forces still play an important role in the local prevention, control and understanding of epidemics.

Tibetan medicine is extensive and profound with its unique system of medicines, drug processing, prescription, preparation, diagnosis and treatment, and boasts a history of centuries in the Tibetan society. The Mongolians also absorbed the local Tibetan medicine after they entered Henan County in the mid 17th century, and developed it into their own traditional medicine. In history, the big temples in Tibetan areas set up "Manbazhacang" (Tibetan medicine institute) to specialize in medical treatment and education. The scale of temples in Huangnan Pastoral Area was small and had no traditional Tibetan medicine institute. But lamas always went to big temples in Tibet and Qinghai for to study Buddhism, almanac, philosophy and medicine, and some lamas specialized in medicine. After finishing their study, many lamas went back to their

[9] S.L.Yue and G.L.Wang. Analyses for the Monitoring Data for Plague in Human Habitation in Qinghai Province. *Chinese Journal of Preclinical Medicine*, 2001, No4.

[10] *Records of Zeku County*. Op.cit., p.464.

hometown, for example, famous Tibetan Doctor Kezeng etc. They all had the experience of studying medicine in the temples in other places. After they finished their study, on one hand, they practiced medicine and continued to study intensively and carefully by combining with the local situations, and on the other hand, they taught their disciples and made their efforts to extend their own knowledge. In the long-term medical practices, many distinguished Tibetan doctors were cultured. Moreover, there are some Tibetan doctors who practice medicine in folk. There were 54 Tibetan doctors in Zeku County in 1928[11], and there were also more than ten Tibetan doctors in Henan County before 1950[12]. After the foundation of the People's Republic of China, Tibetan medicine was gradually included in the national medical and health system, but folk Tibetan doctors still exist. Deputy director of Disease Prevention and Control Center of Henan County has worked in Henan Mongolian and Tibetan Hospital for more than ten years, and he said that there were a lot folk Tibetan doctors in folk and most of them became doctors by teaching themselves and imperceptibly influenced by what they constantly see and hear during their childhood. They take part in the production in normal time and practice medicine when they are asked for a treatment, and the government issues medical license for them.

However, although the medical treatment in Huangnan Pastoral Area mainly depends on Tibetan medicines, the scale is limited. Before the foundation of the People's Republic of China, Tibetan medicines were very expensive in Huangnan pastoral area: a pill of Renqingrili can be exchanged for a horse, which is more than 120 times of 5 Yuan for one pill in 1990[13]. Due to the deficiency in doctors and medicines, the life of herdsmen was poor and they could not see doctors when they fell ill, so they can only entrust their hopes for survival and disasters elimination to the care of deities. The Tibetan and Mongolian in Huangnan Pastoral Area generally believe in Tibetan Buddhism, and this religion assumed that the occurrence of epidemics comes from karma or improper behavior in the reality. People always let epidemics stew in their own juice. When an epidemic was serious, they would go to temples or burn joss sticks, recite scriptures and worship Buddha and the tribe would carry out collective pray activities, some people would move to other places to avoid the plague. For example, by epidemiological investigations, it was found that

[11] Ibid., p.461.
[12] *Records of Henan County*. Op.cit., p.822.
[13] Ibid., p.810.

the suspected epidemic history of plague among human beings in adjacent Tibetan district in Hainan can be traced back to more than 200 years ago. Due to the limitation of historical condition, the people were lacking knowledge of plague situations and poor capability of plague treatment and prevention; farmers and herdsmen could only adopt the method of abandoning and avoiding the patients suffering from plague, resulting in that the death of infected people and the plague stew in its own juice[14]. For another example, the social and historic investigations on the minority groups in Qinghai in 1950s and 1960s found that the Tibetan families that believed in Red Hats in Huangnan and other places always invited Living Buddha or believers to recite or chant scriptures when somebody in the family fell ill, and each tribe or between tribes also regularly carried out congregations to recite or chant scriptures with different scopes at different scales, and the big gathering of the Tongren county was carried out in Heri Tribe, Zeku County in 1954[15]. Guoluo Tibetan have accumulated certain treating experiences for common animal plagues, but they had no other choices but to beg the deities and recite scriptures when large disasters that have serious hazards on livestock took place, such as foot and mouth disease, ox lung plague, murrain and others[16]. When diseases or hazards on human or animals took place, Haixiwangshidaihai Tibetan tribe chose certain time to recite or chant scriptures according to their characteristics[17].

Until now, the religion still plays important role in the epidemic prevention and treatment in Huangnan pastoral area, and temples are not only places for inviting immortals and praying bless, but also places for Tibetan doctors to practice medicine. The chief of Huangnan Civil Affair Bureau said that Longwu Temple produced large amount of Tibetan medicines during the outbreak of SARS in 2003 and handed them to the mass free of charge. At that time, the society was in panic and people went to temples and Tibetan hospitals for Tibetan medicines. The director of managing committee of Heri Temple in Zeku County said that lama would recite scriptures if herdsmen came to beseech Buddha for illness, and the expense was paid at his own will. The master of Henan Veterinary Station

---

[14] B. J. Zhou. Op.cit.

[15] Z.L.Zhang. Changes of Healthy Communication and Chinese Pestilence Prevention and Cure in Hundreds of Years, Peking University Medical Press, Beijing, 2008.

[16] Compiling Group of Qinghai Province, Social and Historic Investigations for Tibetan and Mongolian in Qinghai Province, Xining, 1985, p.82.

[17] Ibid., p.38.

said that some herdsmen also invited Living Buddha to recite scriptures in the animal epidemic prevention and treatment. But the amount of vaccines that were received by the village was not sufficient because some herdsmen withheld information of their animal numbers, so the epidemic prevention workers had to cut down the dosage of vaccine. Some herdsmen thought half a bottle of vaccine could have one half of the effects, and the other half of effects could be achieved by inviting Living Buddha to recite scriptures, and in this way, the entire effects can be achieved.

In the interview, chief of Delong Village, a Mongolian village in Youganning Town, Henan County, who is 53 years old, told us his understanding and recognition on local epidemics as follows:

There were not many infectious diseases, and he can't remember any serious epidemics had ever happened. The old people in the village all said that the government of Qing Dynasty issued an imperial edict and then no epidemic would take place here. Once someone was ill, people would invite a lama to recite or chant scriptures to prevent the disease and beseech medicines from temples or Tibetan medicine hospitals. If it was common cold, herdsmen who had children would go to buy some medicines. There was a special medicine (it was named "Luomajue" in Tibetan) which consists of six or seven medicines. It would be taken after its maker carried out "Jiachi" (reciting or chanting scriptures), and people who can recite or chant scriptures would do it by himself when he took the medicine. In addition, this medicine can also be used to wash a face. This medicine can prevent many kinds of diseases and has good effects. But it would have no effects without reciting scriptures or "Jiachi". Many people took "Luojuema" during the outbreak of SARS in 2003. But there are more animal epidemics in the village and epidemic prevention was done well during collective economy period. In current individual economy period, some people inoculated vaccines for animals and some people did not, some people inoculated more vaccines and some people inoculated less, and thus epidemic prevention can not be carried out uniformly and it was hard to ensure the effects. Ningtemu Town bought a batch of oxen from Linxia last year, which brought foot-and-mouth disease. In the end, 400 oxen were killed at once. As to animal epidemic, it was a tradition to invite Living Buddha to recite scriptures or feed pills to livestock with pills made by themselves, and all the villagers here can make the pills. In the past, there were lots of rabies here. If people or ox was bitten by a dog with rabies, they would suffer the same disease. It was said that wizards (DAWA) can cure rabies and he cut the belly of the patients open and released blood, then sprayed milk on them and drew the milk out. Rabies can be cured after DAWA touched the patients with their hands. These kinds of phenomena no longer exist, and people did not know its reason though they did not inoculate vaccines for their dogs. DAWA had some magic power and they can't be shot with guns.

With the development of economy and society, the understanding on epidemics of the mass in Huangnan Pastoral Area is changing. For example, the chief of Yangmu Village, Zequ Town, Zeku County said that there were many common diseases in the village. Normally, a male patient would go to see a doctor, but a female patient would feel embarrassed to see a doctor. Now people who attend cooperative medical system can go to hospitals with their cards, and thus women go to see doctors more often when they are ill. The sanitary habits of women are also changed. However, some people still hope to cure their diseases by reciting safe and secure scriptures. In the past, venereal disease wasn't taken seriously, but when we interviewed the herdsmen from Henan and Zeku, they indicated that they did not talk much about venereal disease publicly. They said if someone was known to suffer venereal disease, he would surely be looked down upon. This means that the ideas of herdsmen have changed a lot, as compared to the situation several decades ago.

III. The Establishment and Development of Modern Epidemic Prevention and Treatment System

Since 1949 Huangnan Pastoral Area has witnessed great social transformations, which are still going on even today. The social transformations have brought about a series of social, economic and cultural changes that are essential to the epidemic prevention and treatment in this area. Typically, the central government has come to be the leading force for the epidemic prevention and treatment here.

Firstly, the Central Government of China pays much attention to the prevention and treatment of epidemics. Once an epidemic breaks out, the government will mobilize all social forces to control and eliminate it as soon as possible. During the earlier years of the People's Republic of China, the health and epidemic prevention work was at the top agenda of Qinghai province. Health and epidemic prevention teams were sent to the rural and pastoral areas to ensure vaccination, conduct investigations on leprosy and venereal disease, and to offer treatment. The Chinese Ministry of Health also sent the health and epidemic prevention personnel to Qinghai to guide and help with the work in the epidemic foci. Since the reform and opening-up, China has shifted its focus of national economic construction to the central and western regions. Great changes have taken place in the health cause in Qinghai, and the epidemic prevention and treatment has been reinforced and developed. In the interview during our field work, we learnt that the epidemic prevention and treatment work in Zeku County has made remarkable progress in recent years, and the Disease Prevention and Control Center of the county has been awarded a variety of merits for its outstanding achievements in epidemic prevention and control. The present director of Zeku Health Bureau, who used to work in the epidemic prevention station of the county, was once awarded the Title "National Advanced Individual of Endemic Disease Prevention and Treatment" in 1985; and the present director of Zeku Epidemic Prevention and Control Center was also granted the title of "National Advanced Individual of Endemic Disease Prevention and Treatment".

Secondly, the construction of national epidemic prevention and control system is reinforced, and has become integrated with the local medical and health institutions at different levels, including county health bureau, disease prevention and control center of the county, hospitals of the county and townships (towns) as well as village clinics, and rural doctors at the grass-roots level should undertake six tasks: epidemic prevention, maternal and child health, propaganda, clinic, family planning, prevention and treatment of endemic diseases and food intoxication (reporting epidemics), so as to improve the folk disease prevention and control system. By the end of 2006, 12 health institutions have been established at the county or township levels in Zeku County, including county hospital, Tibetan hospital of the county, disease control center of the county, health supervision station and 8 township public health centers; there were 64 village clinics as well. There were 91 professional technical personnel and 64 rural doctors in the medical institutions at the county and township levels.

Table 1: Information on Medical Institutions, Hospital Beds and Health Technical Personnel in Some Years in Zeku County[18]

Year	No of institutions	No of beds	Total staff	Among them: Health Technical Personnel			
				Amount	Senior health staff	Junior health staff	Primary health staff
1953	1		3				3
1957	2	16	20	19		12	7
1965	9	25	42	38	6	25	6
1978	10	53	69	63	12	28	23
1988	11	55	140	126	1	16	25
1995	11	83	136	105			

Table 2: Overview of Medical Institutions in Some Years in Henan County[19]

Year		1952	1958	1965	1978	1985	1990
Institution	The People's Hospital	1	1	1	1	1	1
	Mongolian and Tibetan Hospital					1	1
	Township Health Center		2	4	5	6	6
	Total	1	3	5	6	8	8

18 *Records of Zeku County*. Op.cit., p.459-460.
19 *Records of Henan County*. Op.cit., p.809-810.

Bed	The People's Hospital		8	30	42	64	67
	Mongolian and Tibetan Hospital					30	30
	Township Health Center			10	20	27	34
	Total		18	40	62	91	131
Staff	The People's Hospital	4	21	28	31	49	61
	Mongolian and Tibetan Hospital					22	36
	Township Health Center		8	16	21	46	35
	Others				7	28	37
	Total	4	29	44	59	145	169

Thirdly, relevant laws, regulations, rules and emergency system for diseases prevention have been improved, including the formulation of various emergency plans for epidemics of various levels and report system of infectious disease of different levels. Take the example of Henan County. In order to reinforce the administration of the emergency response for the plague prevention and treatment of the county, improve the prompt response capacity, and reduce the losses incurred by infectious diseases, the "Emergency Plan for the prevention and treatment of the plague in Henan County" was formulated in accordance with the "Law of the People's Republic of China on the Prevention and Treatment of Infectious Diseases" and "Regulations on Preparedness for and Response to Emergent Public Health Hazards". And in order to control and wipe out the serious animal epidemics in the county, the county government formulated the "Emergency Plan for the Control of Serious Animal Plagues in Henan County" in accordance with the "Law of the People's Republic of China on the Epidemic Prevention in Animals", "Implementing the Measures for 'Law of the People's Republic of China on the Epidemic Prevention in Animals' in Qinghai province" and "Emergency Plan For the Control of Serious Animal Epidemics in Huangnan Prefecture" by combining the practical situations in Henan County (Illustration 1).

Besides emergency plans with similar content to those of Henan County, some other emergency plans for epidemics were also formulated in Zeku County. For example, the health department of Zeku County formulated the "Prevention and Treatment Plan for Anthrax in Zeku County," targeted at the zoonotic and highly incidental anthrax, an acute infectious disease induced by Bacillus anthracis in Qinghai province (it is listed as class B infectious disease and required to be reported as class A infectious disease by Law of the People's Republic of China on the Prevention and Treatment of Infectious Diseases). In order to effectively control the outbreak and spread of bacillary dysentery and take effective measures against the epidemics, the health administration of Zeku County also formulated the "Emergency Plan for Bacillary Dysentery in Zeku County".

Fourthly, the effective combination of epidemic prevention, maternal and child healthcare and health education plays a great role in the rural society. In recent years, the medical and health departments in Huangnan Pastoral Area have taken great pains to establish and improve the Health Education and Health Promotion network. For example, they took the occasions of "awareness day" and "awareness month" to organize awareness-raising activities on endemic diseases, planned immunization and tuberculosis in Huangnan Prefecture by hanging banners in the streets, arranging posters, handing out leaflets, using publicity vehicles, posting slogans and pictures, handing out condoms, broadcasting and televising; handing out leaflets in both Tibetan and Chinese and carrying out on-spot consultation by taking the occasion to treating endemic and other infectious diseases; spreading various knowledge on prevention and treatment of infectious diseases by means of visiting Living Buddha; carrying out speeches, carrying out on-spot consultation handing out propaganda materials with words and picture in Tibetan and Chinese by taking the occasion of religious activities and gathering. All these activities have proved highly effective.

Fifthly, a new rural cooperative medical system was established. Since 2005, Huangnan Pastoral Area began to adopt a new cooperative medical system in rural areas, which consists of 3 funds: serious disease fund, medical aid fund for and personal account fund. Qinghai province government requires that from 2007 onward everybody should pay 10 Yuan every year and the government subsides 10 Yuan, and the proportion for medical reimbursement in township, county, prefecture and provincial hospitals were 60%, 50%, 40% and 30% respectively. And from January 1, 2007, an advanced payment system has been implemented such that if a herdsman on the co-op medical system sees a doctor within his county, he only needs to pay 50%. The advantage of this system is that it is more convenient for the herdsmen to see a doctor and they can be compensated soon after that. Since the implementation of the new rural cooperative medical system, the local governments in Huangnan Tibetan Autonomous Prefecture have carried out many publicity and mobilization activities so that 98% residents know about the new cooperative medical system (Illustration 2). Although the annual per capita income in this national poverty-stricken county is only more than 1300 Yuan, nearly all the herdsmen are willing to take part in the new cooperative medical system after the mobilization and publicity. In 2005, 41 468 people took part in the cooperative medical system in the whole county, 75.7% of the total population; and by December 25, 2006, the number had risen to 53 207, up to 94.6% of the total[20]. In Henan County, 95% of the herdsmen had attended the new cooperative medical system by the end of 2006. Relating to this, the chief of Delong village of Henan county said that the villagers are all on the cooperative medical system now, and that the new medical system is well received by the herdsmen.

As can be seen, the new cooperative medical system plays an important role in improving the healthcare of farmers and herdsmen, reducing medical expenses, and lifting them from disease-related poverty. The administrative offices and designated medical institutions at different levels also see to it that the new medical system was implemented properly and smoothly, insomuch that the number of attending farmers and herdsmen grew significantly, and the medical expenses of the mass were substantially reduced. At the same time, the medical institutions at

[20] New Medical Cooperative System Management Office of Zeku County, *Report on New Medical Cooperative System Operation in Zeku County*, December 12, 2006.

different levels also saw a great increase in the person-times of clinic, hospitalization and business revenues.

Finally, to support and develop Tibetan medicines. Tibetan medicine has a long history and is very popular in Qinghai province, along with religious activities. Since the founding of the People's Republic of China, the central government has attached great importance to supporting and developing Tibetan medicines. A large number of Tibetan medical institutions have been established in different places, and many folk Tibetan doctors have been adopted as formal medical workers. For example, the Mongolian and Tibetan Hospital of Henan County was established in 1985 as an outgrowth of the former Tibetan Medicine Department of the People's Hospital of Henan County. Xiangzhagabuzang, a famous Tibetan doctor and director emeritus of Qing Hai Tibetan Medicine Hospital, was appointed as a director emeritus of the hospital, and Jiumeiduoji, a famous Tibetan doctor too, was appointed as a director (Illustration 3). Xiangzhagabuzang was much devoted to the collecting and arranging of traditional Tibetan medicines, and managed to develop 8 kinds of precious Tibetan medicines including Renqingzhangjiao. He himself trained more than 60 folk Tibetan doctors for Henan County, and trained more than 40 Tibetan doctors in collaboration with other famous Tibetan doctors in Qinghai province. He also compiled a medical monograph on Tibetan medicines, hereby contributing a lot to the development of national Mongolian and Tibetan medicines. The Tibetan Medicine Hospital of Zeku County was also established in 1985, drawing heavily on the folk Tibetan doctors. Besides the traditional medicaments, the Tibetan medicine hospitals of the two counties also purchased modern equipments for medicament processing and instruments for diagnosis and treatment; the clinical medicines are largely Tibetan, along with traditional Chinese medicines and western medicines though. Although some towns do not have Tibetan medicine hospitals, they do have Tibetan medical clinics. The township hospitals without Tibetan doctors and rural doctors of village clinics also diagnose and treat patients with the aid of Tibetan medicines.

In recent years, Qinghai province has been making efforts to improve the service capacity of Tibetan medicines so as to develop it as an important means of relieving the difficulties and higher expenses in seeing a doctor. Since 2006, Qinghai province has raised the compensation percentage of Tibetan medicines in the new cooperative medical system by 5%, in the benefit of 3 million farmers and herdsmen in Qinghai province. Taking into account the unique role of Tibetan medicines in the diagnosis and treatment of common and highly incidental diseases, Qinghai province organizes the experts in traditional Chinese medicines and Tibetan medicines to select some popular and easily-adopted technologies of traditional Chinese medicines and Tibetan medicines to be generalized (5-10 items every half a year) to the farming and pastoral area[21]. By supporting and developing traditional Tibetan medicine, China has integrated it into the national public medical and health system, which is adapted to the local social and cultural background. It has not only produced good effects in the pastoral areas in Qinghai province, but has promoted the development of medical and health work here. This is an important factor in the successful establishment and implementation of the epidemic prevention and treatment system in Qinghai pastoral area in the past decades.

In addition, the continuous improvement of the status of public health investment in financial expenditure, improvement of public health system, cultivation of personal sanitary habits and public health morality of all the people, and reinforcement of the introduction of international forces are also integral to the development of epidemic prevention and treatment causes in the Huangnan Pastoral Area.

In summary, since the founding of the People's Republic of China in 1949, Huangnan Pastoral Area has witnessed great changes in the political system and public service system as well as economic and cultural development. Accordingly, the epidemic prevention and treatment of the Tibetan and Mongolian society have made prominent achievements, with increasing participation from the social forces, and a general prevention and treatment system at the grass-roots level.

IV. Problems and Suggestions

Based on the investigations and studies above, we argue that the epidemic prevention and treatment in Huangnan Pastoral Area have made much progress since 1949, and that the spread of large epidemics is basically controlled, but there still remain some problems. In addition to the general problems such as lack of a popular social medical insurance for infectious diseases, inferiority of rural medical infrastructure, inadequacy in staffs, equipments and funds, there are also some unique problems.

---

[21] Qinghai Province Effectively Uses Traditional Tibetan Medicines to Relieve The Problems of Difficulties and High Cost in Seeing Doctors for Farmers And Herdsmen, Chinese Taiwan website (www.chinataiwan.org). October 3, 2006.

Firstly, the herdsmen of Zeku County and Henan County are dispersed in a vast area of high mountains and deep valleys, which brings difficulties to the prevention and treatment of epidemics here. Other problems such as the higher costs of going to the countryside, poor transportation and the difficulty of gathering the herdsmen all exacerbate the difficulty of health education and publicity. The medical and health infrastructure in the countryside are poor, the townships and villages in the two counties are unable to provide online epidemic report, and the medical service quality and level needs to be improved. Still, the village doctors are generally ill paid for their overwork.

Secondly, there exist some difficulties in livestock epidemic prevention. Husbandry is the mainstay for the existence and development of Tibetan and Mongolian herdsmen in Huangnan Pastoral Area. The rate of husbandry development is directly related to the rate of casting off poverty and becoming better off, and the epidemic prevention and control is one of the important means of developing husbandry in a productive and effective way. However, as shown in the practices these years, the prevention and treatment of animal epidemics has played a small role. Often, there are some villagers who cover up the real number of their livestock, which results in the deficiency in vaccine distribution. As a result, some folk epidemic prevention staff has to reduce the injection dosage, hereby affecting the effects of epidemic prevention. Other herdsmen have a rather weak awareness of epidemic prevention, assuming that the part of the effects of epidemic prevention could be achieved by the chanting of a living Buddha. Since the folk epidemic prevention workers are generally not adequately paid for their overwork, their enthusiasm has been dampened. Few folk veterinarians are willing to be epidemic prevention workers, so the epidemic prevention work is almost paralyzed. Consequently, some epidemics that used to be eliminated and controlled recurred in time, which has caused great losses in the husbandry production and economic development. For example, there are 138 folk veterinarians in Henan County. The annual income of a folk epidemic worker is only 360 Yuan, while the number of livestock in the county has increased from 750 000 to 1 million, up by 30%. That is to say, each epidemic worker has to vaccinate 10 000 livestock in a year. At the same time, since no target responsibility system is at work, there have emerged a series of problems, such as lack of responsibility and lower expertise for epidemic prevention workers; the non-standard management and frequent transfer of folk veterinarians, as well as the low educational level of folk veterinarians, which leads directly to disqualified or insufficient training. All these have offset the effects of the epidemic prevention of livestock.

Finally, traditional ideas and customs are still influencing the behaviors and mentality of the people. Since the herdsmen in some areas traditionally hunt, skin and eat marmots, which is likely to trigger the spread of plague. It also interferes the national prevention and control of the plague. Some herdsmen even obstruct plague detection so that the detection cannot be conducted normally. For example, the density of marmots in Tantuhu village, Duohemao town, Zeku County was as high as 1.4 marmots/hectare several years ago, so the county health department had planed to send a plague monitoring team to the village, but the plan was thwarted by local residents. In August 2004, the Disease Prevention and Control Center of the county carried out publicity and education in that county and held a number of talks with the village leaders, to little avail. After a round of negotiations and consultations, it was finally agreed that as many as 50 marmots could be hunted from each village, and that for each marmot hunted, an additional 10 Yuan would be paid to the villagers. But the villagers in Tatuhu Village still would not allow the hunting of marmots, such that the plague monitoring work in not yet in place in that village.

In general, the existing problems of the medical and health care in Huangnan Pastoral Area mainly derive from the prominent contradiction between the current status of public health work and the increasing health demands of the people as well as their adaptation to market economy. Due to the historical, geographical and economic constraints, the health capacity in the pastoral area is still weak, and the health work here needs to be strengthened. Despite the steady progress in the primary medical and health work, the criteria are relatively low, and there remain many difficulties in consolidation and expansion; the infrastructure in the public health centers of the town/township are relatively poor, and the professional level of the staffs is relatively low; and the health work faces the dual burden of infectious diseases and chronic no-infectious diseases. although the morbidity and mortality of infectious diseases, endemic diseases and parasitic diseases have been significantly reduced, the pre-conditions and epidemic foci are not yet removed. In the farming and pastoral areas in particular, the acute and chronic infectious diseases and endemic diseases have not been brought under control, while other emergeng infectious diseases such as venereal disease and AIDS pose new threats.

Based on this understanding, we contend that the following improvements can be made to the epidemic prevention and treatment in Huangnan Pastoral Area: firstly, promote the platform for public health service in the countryside, so as to offer secure, effective, convenient and cheaper public health and medical services. On the one hand, promote the development of the basic medical and health institutions at township and village levels, which is the basic platform of ensuring the effects of basic medical and health projects, and the main carrier of basic medical and health system construction in the countryside. On the other hand, we should improve the staff quality at the positions in the countryside because it is the important guarantee of the basic medical and health service construction in the countryside. At the same time, in order to develop and improve the new cooperative medical system in the countryside, we may exempt the ten Yuan that should be handed over by herdsmen who attend the new cooperative medical system since too much labor and materials are involved to collect the money.

Secondly, the animal epidemic prevention and treatment is an important measure for the development of husbandry. In order to bring the initiative of the folk veterinarians into full play and do well in the domestic animal prevention and treatment at the village level, we may start with the following aspects: a) implementing engagement system for the folk veterinarians in the villages and stabilize the folk veterinarians; b) reinforcing the professional training for folk veterinarians and helping them to learn and master the technologies for the diagnosis of common diseases as well as the prevention and treatment of infectious diseases to ensure their overall level in the villages; c) the work of folk veterinarians should be linked with their payment, and the folk veterinarians at all levels should sign the target responsibility statement agreement with the township veterinary station and local government, such that excellent veterinarians should be awarded and inferior veterinarians should be punished; d) improving the working conditions of folk veterinarians, and their payments should increase in accordance with the charges for epidemic prevention; e) defining the tasks of folk veterinarians, and the veterinary stations at grass-roots level can improve the report system of epidemics according to the work task indexes from the departments at the higher levels, local livestock distribution as well as the spreading situation of epidemics[22].

---

[22] Cairangdangzhi. Several measurements for improving the epidemic prevention for livestock in basic level. *Husbandry in Qinghai,* 2001, No2.

Concluding Remarks

In general, disasters are closely related to epidemics, and it has been a common phenomenon in the world that big plagues always follow big disasters. However, after the investigation, we found that the general epidemic situation of plagues in the Huangnan Pastoral Area, Tibetan Autonomous Prefecture, Qinghai has been improved a lot, although natural and biological disasters still take place occasionally in the past decades. Herdsmen said that the hazards of disasters had been reduced, especially that the death rate was reduced, and thus the risks of epidemics after disasters were also reduced. This indicates that the capabilities of disasters' prevention and reduction in the local Tibetan and Mongolian society have been significantly improved.

The progress in the epidemic prevention and treatment of plagues in Huangnan Pastoral Area is not merely a result of the transformation from spontaneous folk prevention and treatment to that of the central government, but a result of the transformation from treatment to prevention and legalized epidemic prevention. In some sense, it has changed the extent of dependency of the people on traditional social organizations and religions, and reflects the continuous intensification and reinforcement of the central government in remote minority areas. At the same time, China also gradually reinforces the participation of social forces besides constructing epidemic prevention and treatment system of general improvement at grass-roots level. It is the corresponding system guarantees from the country that brings the more effectiveness of social strength, therefore the effective and direct managements from the country and social mobilization are interactive and complementary.

References

Cairangdangzhi. Several measurements for improving the epidemic prevention for livestock in basic level. *Husbandry in Qinghai,* 2001, No2.

China Academy of Chinese Medical Science, Chinese Pestilence History, Chinese Ancient Classics Press, Beijing, 2003.

Compiling Group of Qinghai Province, Social and Historic Investigations for Tibetan and Mongolia In Qinghai Province, the People's Press of Qinghai Province, Xining, 1985.

Li C. N. Epidemiological Investigations on the Venereal Disease among Tibetan in Remote Pastoral Areas in Qinghai Province, *Chinese Epidemiology Journal*, 2004, No3.

Luo S. D. W., C. Li, Z. X. Wang. Analyses for the first case of plague in human habitation in Qinghai highland, *Chinese Journal of Endemiology,* 2002, No3.

New Medical Cooperative System Management Office of Zeku County. Report on New Medical Cooperative System Operation in Zeku County, 2006, Dec. 12.

Records of Zeku County compiled by the Compiling Committee of Records of Zeku County. Chinese Annual Review for Records of County and Town Press, Beijing, 2005.

Records of Henan County (the second part) compiled by the Compiling Committee of Records for Henan autonomic Mongolia County. People's Press of Gansu, Lanzhou, 1996.

Wang H. Z. Disaster, Society and Modernization: Review Focusing on North Jiangsu during the Period of the Republic of China, Social Sciences Academic Press, Beijing, 2005.

Yue S. L., G. L. Wang. Analyses for the Monitoring Data for Plague in Human Habitation in Qinghai Province, *Chinese Journal of Preclinical Medicine*, 2001, No4.

Zhang Z. L. Changes of Healthy Communication and Chinese Pestilence Prevention and Cure in Hundreds of Years, Peking University Medical Press, Beijing, 2008.

Zhou B. J. Review on the fifty years for the prevention and treatment of plague in Huangnan Tibetan Autonomous Prefecture, Qinghai Province, *Bulletin of Endemic Diseases*, 2006, No3.

Zhu J. Q., etc., Investigations on the Characteristics of the Spread of plague In Human Habitation and the Control Measurements in Qinghai Province, *Bulletin of Endemic Disease*, 1993, Vol. 8, No1.

Internet Sources

Qinghai Province Effectively Uses Traditional Tibetan Medicines to Relieve The Problems of Difficulties and High Cost in Seeing Doctors for Farmers And Herdsmen, Chinese Taiwan website (www.chinataiwan.org).

Illustrations

1. Disease Prevention and Control Center of the Henan County

2. Emergency treatment equipment baggage at Disease Prevention and Control Center of the Huangnan Tibetan Autonomous Prefecture

3. Tibetan Medicine Hospital of Henan County

Abstract

The epidemic disease is different from ordinary disease in that it is likely to generate sizable political, economic and cultural problems once it happens. The Huangnan Prefecture in Qinghai Province is a pasture as well as a compact community for Tibetan and Mongolian herdsmen. Fieldwork research revels that some parts in this area were once the source of rat plague, as well as of some other epidemic diseases. The outbreak of these epidemics is closely related to the local natural environment and to the social-cultural practice. The local government used social forces including Tibetan folk medicine, the temple and other regulations to restrain peoples' behavior in daily life to prevent and alleviate the epidemic diseases. After the founding of People's Republic of China, Qinghai pasture has witnessed fundamental changes in its social system, and the relationship between the locality and the state has become more assimilated to that of the interior of China. Like in other walks of life, the state is deepening and reinforcing the efforts to improve the medical and health conditions in Qinghai pasture.

Fang Sumei, Ph.D. in History, Professor (senior research fellow) at the Department of Ethno-History – the Institute of Ethnology and Anthropology to the Chinese Academy of Social Sciences. Research interests are focused on modern history of China's minority nationalities, ethnic issues and ethnic policies in modern China, minority nationalities' social and cultural changes in modern China. Main publications: *A Study of the Zhuangs'Society in Modern Times (2002); A History of Minorities' Revolution in China, 1840-1949 (2000).*

BULGARIAN TRADITIONAL ECOLOGY

TANYA BONEVA
ASSOC. PROFESSOR, DEPARTMENT OF ETHNOLOGY,
SOFIA UNIVERSITY, ST.KLIMENT OHRIDSKI

Bulgarian popular culture, as virtually every type of such culture, produced over the centuries a kind of cultural representation of disasters associated with the natural environment, and also gave a cultural response to them and established some patterns of their prevention. These have come down to us through folklore, remembered in rites and rituals, through fixed or occasional feasts and holidays, descriptions and portrayals, and "embedded in group's collective conscience as permanent markers of social time"[1]. The pages below aim at studying some of these markers as expression of the specific way of Bulgarian traditional culture to deal with nature by successfully adapting to it.

The ethnic territory of the Bulgarians is part of Southeast Europe, populated by humans since the Middle Paleolithic Age and inhabited without any interruption to date. More than 50 settlement mounds, dated back to the 6th-3rd millennium B.C., have been excavated in Bulgaria, in which grain grinders, hand operated farming tools and the like have been found, pointing to the development of agricultural production[2]. Settlement mounds have been uncovered in various parts of the territory of Bulgaria. Most of them are in South and North-East Bulgaria, as well as in the region of Sofia. Best known are those in the lands of the village of Karanovo, Nova Zagora region, the village of Yunatsite, Pazardjik region, the village of Vesselinovo, Yambol Region, the village of Vinitsa, Varna region, and the village of Kremikovtsi, Sofia region. In these settlement

---

[1] A.Eyre, T.Wachtendorf, and G. R. Webb. Introduction."Bringing Culture Back In: Exploring the Cultural Dimensions of Disaster," – In: A.Eyre, T. Wachtendorf, and G. R. Webb (eds.). The Popular Culture of Disaster. – In: University of South California, School of Policy, *Planning and Development*, Issue Vol.18, No1, March 2000, p. 5-19.

[2] R.Dennel. Early Farming in South Bulgaria from 6th to 3rd Millennium B.C., Oxford, 1978.

mounds life continued for 3-4 thousand years. During the Bronze and the Early Iron Age, the ethnic composition of the population of this part of Europe changed, but farming and stock-breeding remained the staple economic occupations.

The Bulgarian lands are part of the zone of Eurasian grain production. The ancient agrarian heritage in the culture which developed in the Bulgarian lands included grain production, as well as viticulture[3]. The iron parts of the tools used by the peasants, and preserved to this day, were made after models developed in Antiquity[4]. The forefathers of the modern Bulgarians – Slavs and Proto-Bulgarians, as well as the different ethnic communities which migrated and settled here, preserved the local heritage and developed complicated mechanisms of adaptation to the types of soils, climatic conditions, altitude, and also to the use of wild growing plants and animals, etc.

A specific feature of traditional Bulgarian ecology has been the development of comparatively stable mechanisms of man's impact on the state of the environment. By means of specialized tools manufactured in the household, knowledge of and skills in growing grain crops and cultivating the fields acquired in the families and local communities, the rural population established harmony with the environment to the end of securing fertility of the soil[5].

Existence for a long time of the same population in a given environment of certain altitude, specific climate, availability or shortage of water, closeness to or great distance from roads and other forms of communications (rivers, seas) also leads to development of specific social forms of adaptation: family, local and religious communities (settlement, municipality) aimed at ensuring the sustainable development of the population in the territory of the settlements or, possibly, under the conditions of migration at certain moment. The political conditions, in which the Bulgarians

---

[3] M.Chukleva. Za antichniya proizhod na dva lozarski sada: chebar I korab. [About the Ancient Origin of Two Viticulture Vessels: *Chebar* and *Korab*] (Using materials from the Burgas District), *Izvestiya na bulgarskite muzei* [News from the Bulgarian Museums], Vol.1, Sofia, 1971, p.212-216; L.Peneva. Lozarstvo [Viticulture] – In: Etnografia na Bulgaria [Ethnography of Bulgaria], Vol.2, Sofia, 1983, p. 26-30.

[4] M.Dukov. Zemedelieto I zheleznite zemedelski oradia prez Antichnostta v bulgarskite zemi [Agriculture and the Iron Farm Tools in Antiquity in the Bulgarian Lands*], Izvestia na Etnografskia institut I muzei* [News of the Ethnographic Institute and Museum], Vol.8, 1965, p.42-58; H.Vakarelski. Etnografia na Bulgaria [Ethnography of Bulgaria], Sofia, 1977, p.91-92.

[5] See: T.Boneva. Ekologia I traditsii [Ecology and Traditions] – In: Kulturna Ekologia [Cultural Ecology] (Collection). Compiled and edited by T.Boneva, Sofia, 1997, p.317-320.

have been preserved as an ethnic formation over the centuries – since the setting up of the state of Bulgaria during the Middle Ages (AD 681) to date – have also influenced the forms of adaptation aimed at containing the economic and demographic crises and catastrophes.

Up to the first half of the 20th century, the Bulgarian peasants lived in one of the three basic agrarian ecosystems: movable (cutting and firing) agriculture, plough agriculture which produced mostly grain crops, and irrigated agriculture[6] connected with the development of rice growing. Alongside the agricultural staple crop-grain, over the past few centuries the production of technical crops has also been developed: roses, cotton, tobacco and others[7]. Nomadic agriculture protected from demographic catastrophes those inhabitants of the mountains who strove to adapt to the maximum use of the raw materials in the lands of their settlement. Nomadic agriculture was part of their ecological niche, including stock-breeding (mostly sheep-breeding). The lowland forests were cut down and the spaces around the forests were cleared where temporary fields were planted and grain was sown in the cinders remaining from the fires. Such fields kept on their fertility for no more than 2 or 3 years[8]. Agriculture by using a wooden plough of the type of the Roman plough used to be the predominant traditional technology of land cultivation[9]. The most widespread agricultural crop was wheat[10], followed by rye and oats. As it has

[6] H.E.Odum. Energy Basis for Man and Nature, New York, 1976, p.140-147.
[7] This was a purposeful policy of the Ottoman Empire which exercised political domination over the Bulgarians from the end of the 14th century till the establishment of the Bulgarian national state in 1878. Since the 18th century onwards rice, cotton, sesame and other crops have been cultivated in the southern parts of Bulgaria (Plovdiv and Haskovo regions), while attar of roses has been produced in the region of the town of Kazanluk (See: Stopanska instoria na Bulgaria [Economic History of Bulgaria], Sofia, 1981, p.183).
[8] In the regions of the Rhodope and Strandja mountains the periodical making of new fields continued to as late as during 1940s. This type of agriculture was described by S.Raichevski. Zemedelie [Agriculture]. – In: Strandja [Collection Strandja]. Sofia, 1996, p.28-29.
[9] The triple system of soil cultivation, whereby two years running, the field was planted, and during the third it was left "to rest", i.e. people relied on the natural restoration of land fertility. (See: H.Vakarelski. Zemedelie [Agriculture]. – In: Etnografia na Bulgaria [Ethnography of Bulgaria], Vol.2, Sofia, 1983, p.15.
[10] The most widespread grain variety is Triticum vulgare with the variety Triticum durum erythrospermum (See: H.Vakarelski. Etnografia na Bulgaria [Ethnography of Bulgaria], Sofia, 1977, p.91-92). Different varieties are known in different regions of the country: *chervenka, zagaria, arnautka* (a variety of hard corn). *Kar-*

been pointed out, some of the farmers in the south part of Bulgaria specialized in the growing of technical crops and rice in the fertile plains of the country. The farmers made great efforts to guarantee the fertility of the wheat fields.

These efforts also included symbolic influencing Nature through the system of cults of Christian saints who, in the holiday system of the farmers, played the role of patron saints of fertility, as well as of demiurges (creators) of techniques and technologies of production. In these cults the Christian canon was reinterpreted to include the idea of fertility of the winter wheat or the other farm crops which the families and the settlements produced in complete dependence on the season and the climatic hazards of infertility caused by droughts, floods, thunders and fires, plant diseases and the like. The system of holidays through which fertility of the winter wheat was influenced conformed to the cycle of its sowing during the autumn (end of September/October), its germination, the plants remaining in the earth during the winter, its development and ripening in the spring (March/May) and harvesting in the summer (end of June/August). For that reason fertility was ensured by a series of rites and rituals to different saints: St Andrew (October 30), St. Ignatius (October 20), Christmas Day (December 25), New Year (St Basil's feast day), St Vlas (February 10), Cheese Fast Sunday and the rituals united by the common name *Kukeri* Masked Dances, Easter, the feast day of St George, St Elijah's feast day (July 20) etc.[11]. On these feast days sets of rites and rituals were performed the purpose of which was ensuring fertility. Of these worth describing is the ritual feast on Christmas Eve. This used to be a family feast prepared in every village house. This was also the last meat-fast meal of the Christmas fasting at which, on Christmas Eve (when the birth of Christ is awaited), the picture of the productiveness of the family was reflected in the dishes served; they were prepared of all the farm crops produced on the family farm, including ritual loaves of bread the decorations on which represented different farming activities (a field, a vineyard, a

---

nobatsko and others. (See: A.Popov, K.Pavlov. Rastenievadstvo [Plant Growing], part 1, Sofia, 1966, p. 50-56).

[11] About this see: M.Vassileva. Kalendarni praznitsi I obichai [Calendar Holidays and Rites and Rituals]. – In: Dobrudja [Collect6ion Dobrudja]. Etnografski, folklorni I ezikovi prouchvania [Ethnographic, Folklore and Linguistic Studies]. Sofia, 1974, p.304-306. M.Vassileva. Kalendarni praznitsi I obichai [Calendar Holidays and Rites and Rituals] – In: Etnografia na Bulgaria [Ethnography of Bulgaria], Vol.3, Sofia, 1985, p.95-127; 216-222; T.Boneva. Ekologia I traditsii [Ecology and Traditions], – In: Kulturna ekologia [Cultural Ecology], Sofia, 1997, p.327-329.

sheepfold) and also domestic animals. In this way, at the moment when the Holy Virgin gives birth to Jesus Christ, the Son of God, a table was laid "for the new mother, the Holy Virgin," the abundant food on the table representing the prosperity of the family. Symbolically represented at this feast arranged on wheat straw and ears of wheat and parts of the plough was the field of the peasant and its expected fertility. The relation of the feast with the moment of birth determines its symbolism: at that moment the fertility of the fields "is born"[12]. Fertility is reaffirmed in this and in a series of similar rites and rituals performed from the moment of birth to the baptism of Christ in the waters of the Jordan River on January 6th (Epiphany) through a repetition of a ritual feast similar to the one described and the addition of various ritual meat dishes.

I would like to emphasize that the Holy Virgin occupies a special place in the ideology of fertility also through the rites, rituals and festivals celebrated in summer, when the wheat harvest has ended and some fruit and vegetables have also been harvested. On the day of the Assumption of the Holy Virgin (August 15th) and the Nativity of the Holy Virgin (December 8th) the harvested and gathered fruit is shown and part of it is donated. In Dobrudja valley on this day women carry to church the first grapes and water melons brought from the fields, as well as some of the first honey collected from the hives[13]. In the region of Strandja Mountain a cock is cooked on August 15th, which the families take out and eat in the field. Wheat from the new harvest, grapes, honey and cheese[14] are taken to church in a number of settlements on these holidays. On the festival of the Nativity of the Virgin women hand out bread and bring raw wheat to church, so that the year may have abundant harvest[15]. On this holiday *kurban* offerings and common feasts are held in the family and for the whole village[16]. These ritual practices reflect the idea that the family must give away as donation part of the produce at the indicated ritual feasts to their neighbours and relatives so that there might be an abundant crop. The traditional Bulgarian villager's philosophy, whereby "in order to have, you have to donate" is noticeable in this practice. For that reason, before consuming the abundant crop gathered (wheat, beans, fruit, vegetables, etc.), the family has to donate some of it.

---

[12] T.Boneva. Ekologia I traditsii…[Ecology and Traditions…], p.329.
[13] M.Vassileva. Op. cit., p.339
[14] S.Grebenarova. Kalendarni obichai I obredi [Calendar Customs and Rites and Rituals.] – In: Strandja [Collection Strandja]. Sofia, 1996. p.347.
[15] Ibid.
[16] M.Vassileva. Kalendarni praznitsi I obichai [Calendar Holidays and Rites and Rituals], - In: Etnografia na Bulgaria [Ethnography of Bulgaria], Vol.3, Sofia, 1983, p.131.

The cults of the saints aimed at enhancing fertility are supplemented by a rigorous regulative framework of man's behaviour the purpose of which is to avert or stop short long drought or torrential rains, ward off thunderstorms[17], a demographic crisis in the family and community: "hvarlyane" [casting off] and adoption of a child[18], ward off epidemic and disease[19] etc.

Stockbreeding, as another economic form, also has had centuries-old traditions. Homer, the ancient Greek poet, called Thrace (the southern part of the present-day Bulgarian ethnic territory) "mother of sheep". The region derives its name from its ancient inhabitants, the Thracians[20]. The subdivisions of stockbreeding: cattle breeding (including horse breeding, buffalo breeding) and sheep breeding (incl. goat breeding) developed in their two varieties – nomadic and settled stockbreeding, its purpose being to supplement the diet of the villagers with meat, milk and dairy products.

---

[17] These are the rituals for rain and for stopping it (Peperuda [Butterfly], German), the calendar holidays Bartholomew and Elijah, the Thursdays between Easter and Ascension Day (40 days after Easter), and some others. (S.Genchev. Obichai I obredi za dajd [Customs and Rites and Rituals for Rain], – In: Dobrudja [Collection Dobrudja]. Etnografiski, folklorni i ezikovi prouchvania [Ethnographic, Folklore and Linguistic Studies]. Sofia, BAS, 1974, p.346-351; M.Vassileva. Kalendarni praznitsi I obichai [Calendar Holidays and Rites and Rituals] – In: Etnografia na Bulgaria [Ethnography of Bulgaria], Vol.3, Sofia, 1983, p.110-132.)

[18] Child adoption in traditional Bulgarian culture has been studied in detail and in its various aspects by E.Tzaneva. (See: E.Tzaneva. Edna starinna praktika pri osinovyavane na dete [A Child Adoption Practice of Old Times], *Bulgarska Etnografia [Bulgarian Ethnography]*, 1990, No2, p.23-30 and the references quoted there; E.Tzaneva, Osinovyavane, bibleiski etos I zakon [Adoption, Biblical Ethos and Law], *BE*, 2003, No2-3, p.30-42; E.Tzaneva, Antropologichni teorii za osinovyavaneto [Anthropological Theories Concerning Adoption], *BE*, 1996, No2, p. 51-64; E.Tzaneva. Simvolikata na ritualite pri osinovyavane na dete. [Symbolism and Rituals in Child Adoption] – In: Etnografski problemi na narodnata duhovna kultura [Ethnographic Problems of Folk Spiritual Culture], Vol.2, Sofia, 1994, p. 25-34.

[19] Quite numerous are the practices aiming at warding off disease, which involve exerting magic impact through incantation, "feeding" (laying a table) for the diseases during an epidemic and, as a final measure, leaving the settlements: (Etnografia na Bulgaria [Ethnography of Bulgaria], Vol.3, Sofia, 1985, p.80-88).

[20] On this matter See: V.Marinov. Za etnokulturnite prinosi v balgarskoto zhivotnovadstvo [About the Ethnic and Cultural Contributions to Bulgarian Stockbreeding] – In: Etnogenezis I etnokulturno nasledstvo na bulgarskia narod [Ethno-Genesis and Ethno-Cultural Heritage of the Bulgarian People]. Sofia, 1971, p. 87-88. The names of the farm animals are Slav-Bulgarian (H.Vakarelski. Etnografia na Bulgaria [Ethnography of Bulgaria], Sofia, 1977, p.152-153).

Poultry breeding has also been part of the household farm and has been universally present in the village household[21].

Typical nomadic stockbreeding existed as a system under the form of seasonal migration of the herds of sheep (and goats) between the summertime and wintertime grazing lands and was widespread in all big mountains and plains on the Balkans. In Bulgaria it was connected with two small Christian nomadic communities, Wallachians and Karakachans, which had no permanent settlements and practiced transhumance sheep breeding on the Balkans until the mid-20th century. Producers of meat, wool and diary products, they had a specific economic status and for centuries on end migrated with their herds along established routes[22].

Stockbreeding was the main economic activity of some of the Bulgarians living in the country's mountains. Considerable part of the male population of the villages and small towns in the Balkan Mountain Range, the Rhodopes, Mount Pirin and Strandja, made up a stable community of quasi-nomadic sheep-breeders (sheep-breeders and *kehaya* chief shepherds), whose occupation it was to organize large-scale transhumance sheep breeding, following the nomadic seasonal cycle of raising the herds in high mountain pastures during "the winter" and in low plain pastures during the "summer" in conformity to the life cycle of the animals. "The winter" in the plain both for the nomads (Wallachians and Karakachani) and for the quasi-nomads covered the period between the feast day of St Demetrius (October 26th) and the feast day of St George (May 6th), when the sheep are inseminated and, respectively, the offspring is born and grows strong; whereas "the summer" (from the feast day of St George /May 6th/ through the Assumption of the Holy Virgin /August 15the/ to the feast day of St Demetrius /October 26th/) is the time of breeding the stock and production of dairy produce in high mountain pastures. The shepherds and chief shepherds spent no more than 2 or 3 months in their native places, but their homes and families lived there. This activity was particularly hazardous because it brought in higher profit, but also caused great losses, should a disease broke out among the animals, or the tem-

[21] I.Zlatarov. Ptitsevadstvoto v Bulgaria [Poultry-Breeding in Bulgaria], Journal of the Bulgarian Economic Society, 1903, No6.

[22] The community of Karakachani was typically nomadic until the end of the 1950s. Some of the Wallachians settled, but those of them who persisted as traditional nomads had no permanent settlements until the end of the 1950s. (See: I.Georgieva (ed.). Armanite v Bulgaria [The Aromanians in Bulgaria]. Sofia, 1998; Zh.Pimpireva. Karakachanite v Bulgaria [The Karakachani in Bulgaria], Sofia, 1998).

perature dropped sharply, as well as if thefts and incidents occurred on the migration routes which could completely ruin shepherds' livelihood[23].

Characteristic of the local communities producing shepherds was that these were comparatively big settlements with an intricate economic composition, in which men were not only sheep breeders, but also building workers, homespun cloth dress-makers, tailors, fishermen, etc. They also practiced their trades under the conditions of trade migration and worked in different faraway settlements most of the year[24].

Preserving the herds and the life of the sheep during dangerous epidemics was "great skill" and included detailed knowledge of the anatomy and physiology of the animals, methods of treating broken legs, diseases, etc. This knowledge was reflected in the stories of old shepherds recorded in the 19th and early 20th centuries[25]. Their purpose was to provide the best conditions for breeding the animals of different age, to ward off or avert possible crises in their reproduction, to treat diseases and prevent the break outs of epidemics[26]. Treatment by "live" ("God's") fire was applied in the most critical situations – when foot-and-mouth disease, measles and smallpox or some other disease threatened to cause large-scale death of

---

[23] A.Primovski. Bit i kultura na rodopskite bulgari [Lifestyle and Culture of the Rhodope Bulgarians], SbNU, No.54, Sofia, 1973, p.315-317; K.Kanev. Minaloto na selo Momchilavtsi [The History of the Village of Momchilavtsi], Sofia, 1975, p. 230-236; T.Boneva. Profesionalni groupi [Professional Groups] – In: Rhodope [Collection Rhodope Mountains], Sofia, 1994, p.235-236; L.Dukov. Zhivotnovadstvo. [Stockbreeding] – In: Dobrudja [Collection Dobrudja]. Etnografiski, folklorni i ezikovi prouchvania [Ethnographic, Folklore and Linguistic Studies], Sofia, 1974, p.230-249.

[24] H.Vakarelski. Etnografia na Bulgaria [Ethnography of Bulgaria], Sofia, 1977, p. 35-46; V.Marinov. Ovtsevadstvo i kozevadstro. [Sheep-Breeding and Goats-Breeding], - In: Etnografia na Bulgaria [Ethnography of Bulgaria], Vol.2, Sofia, 1983, p.43-50; K.Kanev. K. Minaloto na selo Momchilovtsi [The Past of the Village of Momchilavtsi], Sofia, 1974; L.Dukov. Kotlenskoto ovtsevadstvo v Dobrudja [Kotel Sheep-Breeding in the Dobrudja] (In L.Dukov, Zhivotnovadstvo [Stock-Breeding] – In: Dobrudja [Collection Dobrudja], Etnografiski, folklorni i ezikovi prouchvania [Ethnographic, Folklore and Linguistic Studies]. Sofia, 1974, p.73-75; T.Boneva. Profesionalni grupi [Professional Groups], – In: Rhodope [Collection Rhodope Mountains] Sofia, 1996, p.230-249; S.Rakshieva. Zhivotnovadstvo [Stock-Breeding] – In: Loveshki Krai [Collection Lovech Region], Sofia, 1999, p.31-60.

[25] S.Shishkov. Zhivotnite v svetogleda na rodopskite poseleniya [Animals in the World View of the Rhodope Settlements], *Rodopski napredak* [Rhodope Progress], 1912, NoNo4, 5, 7, 10; S.Rakshieva. Op. cit. p.52-54.

[26] See: reference 23 in the studies of V.Marinov, S.Rakshieva, L.Dukov.

sheep – whereby the sheep were taken through or over "live fires"[27] lit by rubbing two pieces of wood together. Such knowledge about the domesticated animals was also common among the settled farmers who bred sheep, cows, oxen and other animals on their farms and organized stockbreeding in the form common herds, providing one another with mutual assistance[28]. The fertility and good health of the animals was achieved by the stockbreeders through practicing the cults of certain saints: St George, St Modest, St Petka and others.[29] These included also making offerings of animals (usually a male lamb) and cooking and offering ritual food (*kurban*) at common feasts out of the village or on a certain place (in the Christian temple, at some central place) within its boundaries.

This exceptional mobility of the Bulgarians from the mountainous regions supplied one of the mechanisms for survival of this population which lived in extremely hard conditions, the soils being unsuitable for the development of agriculture, and the exploitation of the resources of the forests, viz. woodcutting, collecting fallen trees and wood processing, etc being also insufficient for its demographic reproduction. Migration became the only way of avoiding demographic catastrophe. The community of the Bulgarian Muslims in the Rhodopes did not take part in the migrations of their Christian Bulgarian neighbours. They survived in the said conditions through stationary sheep-breeding, migrant agriculture and exploitation of the local natural resources.

A specific feature of the development of land property in Bulgaria was the prevalence of petty ownership at the end of the 19th century. Then a trend was noticeable of gradual enlargement of the small-size farms during the first half of the 20th century. In 1946, 37.5% of the farmers owned up to 5 ha of land, and 39% – up to 10 ha[30]. In the period between 1920 and 1944 the farmer population amounted to 5 million and most of them did not find self-realization outside the villages. They could not find employment on the labor market in town, because the industrial development was slower than the natural reproduction of the rural population. A study on

---

[27] About the "live fire" see: S.Shishkov. Sledi ot kulta na oganya v Rodopite [Traces of the Cult of Fire in the Rhodope Mountains]. *Rodopski napredak [Rhodope Progress],* 1909-1910, No5, p.5-11; Etnografia na Bulgaria [Ethnography of Bulgaria], Vol.3, Sofia, 1985, p.222-224.

[28] K.Krastanova. Traditsii na trudova vzaimopomosht v balgarskoto selo. [Traditions of Mutual Assistance in Work in Bulgarian Villages.]. Sofia, 1986.

[29] Etnografia na Bulgaria [Ethnography of Bulgaria], Vol.3, Sofia, 1985, p.222-224.

[30] Stopanska istoria na Bulgaria [Economic History of Bulgaria], 681-1981, Sofia, 1981, p.373.

the state of manpower estimates that in the 1930s the village overpopulation and the "excessive" rural population amounted to 1 million people[31]. The basic social characteristics of the Bulgarian peasants determining the mechanisms of survival and escape from economic, demographic and social catastrophes, was the prevalence of petty land owners in the villages[32].

The development of the economic life and cultural traditions greatly depended on the state of marriage and the family. Marriage and family were a natural and obligatory mode of man's existence ("a man with no family is a half man"). The Bulgarians followed the Christian norms of contraction of marriage: religious endogamy, monogamy, canonical ban on divorce. A specific type of marriage on the Balkans was a norm observed by the Bulgarian peasants almost up to the middle of the 20th century. Monogamy and avoidance of divorce were typical of the behavior in the Bulgarian village until the early decades of the 20th century. Setting up of a family was only possible if strict rules were observed: pre-nuptial virginity and a radical ban on the birth and upbringing of extramarital children, early age for getting married (17-21-22 for girls and 17-20-27 for young men), contraction of marriage within certain age and social groups (between young men and girls, between widowers and widows), subordination to the parents and the family[33]. The social realization of both men

---

[31] A..Totev. Sastoyanieto na prenaselenost v nasheto zemedelsko stopanstvo [Overpopulation in Bulgarian Agriculture], *Spisanie na ikonomichesko druzhestvo [Journal of Economic Society],* XXXIX, 1940, No6, p.353-384.

[32] During the first half of the 20th century and just between the First and the Second World War, 90% of the farms in Bulgaria were defined as small (70%) and "medium-sized" (20%), whereby the main criterion for this categorization was that they did not use hired labour (L.Berov. Sotsialna struktura na seloto v balkanskite strani prez perioda mezhdu dvete svetovni voini [Social Structure of the Village in the Balkan Countries in the Period Between the Two World Wars] Works of the Karl Marx Higher Institute of Economics, Vol.4, Sofia, 1977, p.43-76).

[33] The earliest nuptial age at the end of the 19th century was registered among the Bulgarian Muslims (S.Shishkov. Bulgaro-mohamedanite (pomatsi) Istoriko-zemepisen i nauchen pregled s obrazi, [Bulgarian Muslims (Pomaks), Historical and Local History Scientific Review with Images], Plovdiv, 1936, p.35). In the region of the Pirin Mountain women married at the age of 16-22, and men – between 18 and 25. The poorest (hired hands) marries later – between the ages 25 and 30. (I.Georgieva. Obichai pri svatba [Marriage Customs] – In: Pirinski kray. Etnografski, folklorni I ezikovi prouchvaniya [Pirin Region. Ethnographic, Folklore and Linguistic Studies], Sofia, 1980, p.390). The Age of Marriage of Girls in the Strandja – 17-22 Years, and of the Young Men – 20-25 Years: T.Boneva. Sotsialni grupi I obshnosti…. [Social Groups and Communities….], p.367; Etnografia na Bulgaria [Ethnography of Bulgaria], Vol.3, Sofia, 1980, p 95-132, 335-341.

and women was only possible through the family. The village circle, also called *zadruga* [patriarchal community], combined agriculture and settled stockbreeding. It included one or more families, established according to the principle of male kinship. This family relied on the natural reproduction to increase the manpower, and included various forms of exchange of labor (mutual assistance[34]) with other families. The family household organized the economic and social activities of the family through common work, common life in a common house[35], common meals, common upbringing of the children, selection of marriage partners for the young, preparation of the personal property (dowry) of the daughters (excluding real estate) for their marriage, building of a house (houses) for the sons, organization of weddings and a number of family and calendar rites and rituals etc.[36] The family household periodically split into separate families which started independent economic and social activity. The family's social relationships with the other families and the local community as a whole were maintained by the oldest man, helped by his wife. Individualism was not part of the upbringing of the children and the young people in the rural family described. Having reached marriageable age, the sons got married and lived (along with their families), sometimes for several years, together with the family of their father, taking part in the collective work in agriculture and stockbreeding. These men reached their complete social realization relatively late in life, after their father had built separate houses for them and had given them part of the arable land (the fields) to use. Such independence was possible to reach only after the death of the father

---

[34] See: footnote 23.

[35] In certain occasions individual families reside in separate homes, which are inadequate in structure (without a fireplace). On this matter see: Tz. Manova. Pomeshtenia za novobrachni v Pernishko v kraya na XIX – nachaloto na XX vek [Premises for Newly Married in the Region of Pernik, at the End of the 19th-Early 20th Century]. *Bulgarska etnografia* [Bulgarian Ethnography], No2, 1975, p.50-59.

[36] About the family see: R.Peshev. Semeistvoto i negovoto istorichesko razvitie [Families and Their Historical Development] – In: Etnografia na Bulgaria [Ethnography of Bulgaria], Vol.1, Sofia, 1980, p.286-302; L.Makaveeva. Bulgarskoto semeistvo: Etnosotsialni aspekti, [Bulgarian Families: Ethno-Social Aspects], Sofia, 1991; M.Todorova. Balkan Family Structure and European Patterns: Demographic Development in Ottoman Bulgaria, Washington, 1993; M.Todorova. Balkanskoto semeistvo: Istoricheska demografia na bulgarskoto obshtestvo prez Osmanskia period [The Balkan Family: Historical Demography of Bulgarian Society in the Ottoman Period], Sofia, 2002; Tz.Boncheva. Brak i semeistvo pri katolitsite v Plovdivsko prez parvata polovina na XX vek [Marriage and Family Among the Catholics in the Region of Plovdiv During the First Half of the 20th Century] Sofia, 2005.

– the oldest one – in which case the property of the family was divided among all the sons. It is noteworthy that traditional law deprived the daughters in the family from rights to real estate. Should a bride receive a field from her father as dowry, this was considered a sign that she had broken the norms of behavior established by the Christian canon and the rules of conduct of local tradition: she had lost her virginity and there was a risk of dissolving the marriage. This would have been a catastrophe for the fate of the young woman and which was prevented through breaking the traditional rules for property inheritance in the family[37]. In the mountainous parts of Bulgaria, where major occupation of the men was nomadic sheep-breeding, as well as different non-agricultural trades connected with manufacturing of goods for the market and the employment of hired labour, the young families, more often than not, separated from their elders. The age of contracting marriage in these cases was higher (25-27 years) and was reached when the men had acquired skill in trade and social prosperity. Social independence was gained through man's self-realization in his chosen trade[38].

Demographic stability and sustainable development of the settlements as communities was achieved by following the traditionally established custom of marriage endogamy in the settlement (for the bigger settlements) or clearly formulated rules of nuptial exchange between certain smaller settlements. This traditional rule was determined to a great extent by both the endogamous religious character of the Christian marriage and the specific political and religious conditions of the Christian peoples on the Balkans which lived under the rule of the Islamic Ottoman Empire in the period between the 15th and 19th centuries. As part of the non-Muslim population of the Empire, the Bulgarians ("the infidels", as they were called by the Muslims) had limited social rights and the vast majority of them survived in their closed local communities. Another, much smaller part of them, adopted Islam as a way of integration into the Muslim environment and in search of a possibility for social prosperity[39].

---

[37] P.Hristov. Sotsialna marginalizatsia i narodna kultura. [Social Marginalization and Folk Culture] – In: Etnografski etyudi. [Ethnographic Studies], Sofia, 2006, p.39-40.

[38] M. Todorova. Balkanskoto semeistvo, [The Family on the Balkans], Sofia, 2002, p. 44; T.Boneva. Esnafski organizatsii, [Professional Organizations] – In: Rhodope [Collection Rhodope Mountains], Sofia, 1996, p.264; T.Boneva. Sotsialni grupi i obshnotsti [Social Groups and Communities] – In: Strandja [Collection Strandja], p.357-360.

[39] S.Raichevski. Bulgarite mohamedani [Bulgarian Muslims] Sofia, 1998, 2nd edition 2004; Tz. Georgieva. Pomatsi – Bulgari myusyulmani [Pomaks – Bulgarian

The settlement endogamy among the Christian Bulgarians did not preclude, but restricted marriages in other settlements and regulated the nuptial exchange with the purpose of preservation the demographic wholeness and ethnic character of the local communities[40]. For the Bulgarians, their Christian faith had ethnic character (it was "a Bulgarian faith"). This mechanism of securing demographic stability of the local communities was preserved in the country until almost the middle of the 20th century and shaped the value system of the peasants during the period. This was marriage to "one of ours"[41], life within the framework of "one's own" local and ethnic community, work as a means of survival and acquiring wealth ("work for no purpose, but do not stay idle for no purpose"), social

Muslims.] – In: Obshnosti I identichnosti v Bulgaria. [Communities and Identities in Bulgaria] Sofia, 1998, p.286-308. A.Zhelyazkova. Turtsi. [Turks.] – In: A.Krasteva (ed.). Obshnosti I identichnosti v Bulgaria. [Communities and Identities in Bulgaria]. Sofia, 1998, p.371-397.

[40] About the local endogamy in the Bulgarian villages see: I.Sanders. Snosheniyata na edno bulgarsko selo s okolnata Sreda. [Relations of a Bulgarian Village with the Environment] – *Zemedelsko-stopanski vaprosi* [Agricultural Problems], 1935-1936, No.3, p.134; V.Filcheva. Semeistvoto v Godechko [Families in the Region of Godech], (A diploma paper at the Ethnography Department of the History Faculty of Sofia University St Kliment Ohridski), Sofia, 1984, p.38-39; T.Boneva. Sotsialni grupi i obshnosti [Social Groups and Communities] – In: Strandja [Collection Strandja], Sofia, 1996, p.361-362; T.Boneva. Etnicheski traditsii v lokalnite obshnosti [Ethnic Traditions in the Local Communities] – In: Izsledvaniya v chest na chlen korespondent prof. Strashimir Dimitrov [Studies in Commemoration of Corresponding Member Professor Strashimir Dimitrov] *Studia Balkanika*. 23, p.763-764; In the studies of P.Hristov on Central West Bulgaria, the model of nuptial endogamy is presented in several settlements in the region of Breznik: P.Hristov. Predbrachno obshtuvane v selo Noevtsi, Breznishko prez 20-te – 30-te godini na XX vek [Prenuptial Communicating in the Village of Noevtsi, Breznik Region in the 1920s-1930s], *Bulgarska etnografia* [Bulgarian Ethnography], 1990, No6, p.9-10. Part of the young people and men and women, who were marginal socially – the poorest, the sick and with special physical marks – were allowed to marry in another village; so were the "old" maidens and bachelors who had crossed the commonly accepted age limit for getting married, as well as widowers and widows. Leaving the village was necessary which was also accepted as a punishment for violating the rules of conduct: T.Boneva. Sotsialni groupi iobshnosti [Social Groups and Communities] – In: Strandja [Collection Strandja], p.360-362.

[41] "A Stranger" was the Muslim, called "a Turk" by the Christians on the Balkans. In folk songs and legends, the Turk was "of another faith"; he was not a suitable marriage partner. About this See: M.Karamihova. Uzhasniyat "obraz" na turchina kato brachen partnyor. [The Horrible "Image" of the Turk as a Marriage Partner.] – In: Predstavata za drugia na Balkanite [The Conception of the Other in the Balkans]. 1995, p.263-267.

solidarity and an egalitarian system of values, in which wealth was valued and social recognition was gained only when attained through working hard.

Settlement's stability as a social community contributed to the gradual evolvement of specific features of economic specialization and discrete specificities distinguishable by its residents in culture (local speech included), setting it apart, to a certain extent, from the other settlements and turning it into a cultural community[42]. The intensity of the communication between settlements based on the closeness of the economy and culture (language included) created regional community and regional specificities in the culture of the Bulgarians[43]. These specificities were the source of internal diversity of the Bulgarian culture, having common cultural system and ethnographic unity[44].

As it has been pointed out, fertility and harmony with Nature were acquired through direct involvement of people. Family was the place where people adopted the ideology of fertility. Within the common Christian rules of piety and devotion, and with the purpose of avoiding sin, specific rules of conduct, bans and punishments were included, which were directly related to fertility. In case of a long drought or rains and floods, the first explanation about the calamity was the sin of a maiden. "God sent" the most horrible calamity – a thunderstorm with torrential rain when "a maiden" gave birth to a child. Then He sent torrential rains "to wash the bad thing". The personal sin of a maiden or a woman – the birth of illegitimate children or children of extramarital relations – brought punishment to the entire community. The impossibility of the family and community to accept childbirth by a maiden entailed the only case of infanticide in traditional Bulgarian society: the murder of a child born "out of

---

[42] This feature in the conduct of the Bulgarians stands out in emigration, too. The local principle of preservation of the Bulgarian culture has been handed down to the young generations and has preserved the Bulgarians in what is today Moldova and the Ukraine, as well as the Bulgarian Catholics in Banat (present-day Romania and individual settlements in Serbia). About this see: T.Boneva. Etnichni traditsii v lokalnite obshnosti. Izsledvane v chest na chl. kor. Prof. D-r Strashimir Dimitrov, [Ethnic Traditions in the Local Communities, Studies in Commemoration of Corresponding Member of the Academy Professor Strashimir Dimitrov, PhD] *Studia Balkanika,* 23, Vol.2, 2001, p.761-783.

[43] S.Genchev. Narodna kultura i etnografia [Folk Culture and Ethnography], Sofia, 1984.

[44] I.Georgieva. Etnografsko edinstvo na svatbata u bulgarite. [Ethnographic Unity of the Wedding Among Bulgarians.] – In: Etnogenezis I kulturno nasledstvo na bulgarskia narod [Ethnic Genesis and Cultural Heritage of the Bulgarian People] Sofia, 1971, p.104-107.

wedlock". The birth of such a child placed it in an unequal position. The child carries a different name, indicating a digression in the conduct of the parents: Ranka (born early), Barzinka (rapidly born), Lefter (a bachelor in Greek), Lefterka (a maiden in Greek) etc., and those children did not have a position of the same standing as the rest of the children[45]. The birth of a child with physical or psychic handicaps was also a catastrophe, which in the traditional conceptions was caused by a violation of a sexual ban (conception on the eve of a major Christian holiday) or came as a result of a sexual contact with a mythical or demonic being (a dragon)[46].

The children born in wedlock had a clear social standing and carried names, established in the family. By way of the Christian baptism, the child received its personal name, which repeated the name of its father's (mother's) father or mother and in this way a stable family anthroponymy took shape across the generations, indicating stability in the reproduction of the family.[47] When a demographic crisis occurred in the family – a premature birth or death of a newborn – it was registered in the system of names in the family through a group of names different from the usual ones, but containing a specific symbolic meaning: Stoyan (deriving from the verb: *stoi,* meaning remain alive), Trayana (to endure, not to die), Darin, Darina, Bozhidar (a gift, a gift from God) or names, connecting the child with the "wild" world – names of birds or animals: Slaveikov (*slavei* – nightingale), Valcho, Valchan (*vulk* – wolf) and others. These digressions in the naming of children indicate the attempts at exerting magic influence on the health and life of the child through giving them names the meaning of which suggested endurance, supernatural help (a gift by God), or else related the child with something alien to the human nature. This traditional mechanism of name-giving points to critical moments when individual families were in danger of extinction, as well as to complex efforts to overcome the threat.

A real picture of the physical standard of life of the Bulgarian peasants can be found in a study of the rural population in the 1930s. It was based on 1420 households in 199 villages in different regions of the country by way of inquiries on their actual state. According to it, the number of the members of a household ranged from 2 to 15, but sporadically there were households enumerating more than 20 members. The average land property owned by a household was 6.05 ha (according to a 1934 population

---

[45] T.Boneva. Naroden svetogled, [Traditional World View] – In: Rhodope [Collection Rhodopes], Sofia, 1994, p.22.

[46] Ibid.

[47] About personal names see: E. Krasteva-Blagoeva. Lichnoto ime v bulgarskata traditsia [The Personal Name in Bulgarian Tradition]. Sofia, 1999.

census, the size of an average farm amounted to 5.22 ha). The same survey showed that 24% of the households listed non-agrarian activities in the set of activities performed on the farm. It was indicative that 63.435% of the basic expenditures of a household were made on food, 9.93% – on clothing, 4.5% – on heating, 1.36% – on lighting etc. The maintenance of the household depended on what was produced in the household's own farm: foodstuffs, materials for clothing, for heating. The household produces 70% of the products they consumed, whereas only 30% of the products were purchased on the market. The main products included in the peasants' diet were: bread (63%), mutton, pork or poultry meat (13.70%), milk and dairy products – 6.84%. It was established that an adult man consumed 3408 calories daily, whereas the energy put in the work amounted to 1800 calories. The daily need of calories for physical work was between 3500 and 5000 calories. Bread had the highest contribution to the energy content of the food (72%), followed by meat and fats (10%), dairy products – 6%, while sugar contributed only 1%[48]. This survey confirmed the importance of grain production and the considerably small importance of meat on the farms.

The bread-other foodstuff ratio attests not only to the importance of grain production for the physical survival of the peasants. The significance of this main food and the much lesser importance of meat and dairy products in the diet of the peasants were also determined by the specific nutrition system enforced by Christian religion. As Eastern Orthodox Christians, the Bulgarians strictly followed the alteration of periods of consumption of food of both plant and animal origin with periods of fasting during which they refrained from eating meat, milk and dairy products, fish and eggs. Fasting extended over a few long periods preceding certain Christian holidays: Christmas fasting preceding Christmas (December 25) lasted for 40 days, Easter fasting preceding Easter – for 49 days, the 15 days preceding the feast day of St Peter (July 29), and the 15 days preceding the Assumption of the Holy Virgin (August 15) were also periods of fasting. Coinciding with the Christmas fasting was the insemination and pregnancy of the female sheep, which were traditionally inseminated on the feast day of St Petka (October 14th) and gave birth at the end of January and in February[49]. The ban on eating meat coincided also with the end

[48] H. Mocheva. Selskoto zemedelsko domakinstvo v Bulgaria prez 1935-1936 g. [The Village Agricultural Household in Bulgaria in 1935-1936]. Sofia, 1938, p.12-14, 22, 40-41.

[49] L.Radeva. Traditsionna narodna hrana [Traditional Bulgarian Food]. – Etnografia na Bulgaria [Ethnography of Bulgaria], Vol.2, p.291-295; L.Radeva. Hranene pri bulgarite [Bulgarian Nutrition] – Etnografia na Bulgaria [Ethnography of

of the active part of the agricultural year and made possible the rest from hard physical work. In this way, at the end of autumn and in early winter, the physical rest was combined with consumption of lower-calorie food lacking in animal fats and proteins, which stimulated the processes of clearing the human organism of toxins. The consumption of food at Christmas (December 25) was associated with ritual slaughtering of a pig and cooking of meals of particularly high caloric contents made out of its meat and fat, rich in cholesterol. This diet coincided with the cold period of the year and supplied energy and warmth. The intake of animal products was again restricted around the end of February when the great fasting began. After its end, on Easter, Bulgarian tradition prescribes the consumption of various kinds of meat dishes: lamb, alongside milk and dairy products. Nutrition is regulated once more after the end of fasting which preceded the feast day of St Peter, when a young chicken is the preferred sacrificial offering to the saint. Chicken meat remains the main food also throughout the autumn and until the beginning of the fasting preceding Christmas. It should be pointed out that, according to the prescriptions of the Christian religion, two day of the week are reserved for fasting throughout the whole year – Wednesday and Friday – and in this way the religious rule considerably restricts the possible consumption of meat and dairy products. Fasting was a time of restrictions not only on the consumption of food; there are no festivals or merry-making of any kind, and there is also a ban on the sexual contacts. These were periods of prayer and trials which aimed at overcoming the temptations of everyday life and sin through restricting the wish for good food, merry-making and reproduction. According to the conceptions of the Bulgarian peasants, quite a number of problems were caused by breaking the norms of conduct during fasting: the conception of a child during this period might, for example, allegedly resulted in the birth of handicapped children. This was a real catastrophe for the family in which such a child was born[50]. In this way, the individual behaviour was regulated through "the supernatural" coercion of the canon and the fear and uncertainty caused by lack of knowledge concerning the causes of natural calamities, disease and abnormalities in the humans. The strict observance of the established norms of conduct in the family and in the local community based on Christian ethics, as well as the regular performance of rites and rituals for land and animal

---

Bulgaria], Vol.II, Sofia, 1983, p.297-298; M.Markova. Traditsionnata hrana i tipologiyata zemedeltsi-skotovadtsi [Traditional Food and Agricultural Workers and Stock-Breeders' Typology]. *BE,* XXXII /2006/, No2, p.161-174.

[50] Personal field research materials from different settlements in the Rhodopes and Strandja.

fertility, as part of the cults for different Christian saints, ensured stability and harmony between man and Nature, whereas the violation of the rules and the break of the symbolic connection with the Christian saints entailed chaos and social and natural catastrophes.

Sustainable economic, social and cultural development was achieved through different forms of adaptation of the population of the individual settlements and regions in the framework of agriculture and stockbreeding, as well as additional (which in many mountainous settlements were also basic) strategies for survival through migration.

The presented traditional system of adaptation of the Bulgarian peasants to the natural environment disintegrated in the middle of the 20th century, when the political and economic system of socialism was established in Bulgaria and private agriculture and traditional nomadic, quasi-nomadic and settled stockbreeding were destroyed.

References

Berov, L. Sotsialna struktura na seloto v Balkanskite strani prez perioda mezhdu dvete svetovni voini [Social Structure of the Village in the Balkan Countries in the Period Between the Two World Wars.]. Works of the K. Marx Higher Institute of Economics, Vol.4, Sofia, 1977, p.43-76.

Boneva, T. Naroden svetogled [Popular World Outlook] – In: Rodopi [Collection Rhodope Mountains], Sofia, BAS, 1994, p.7-50.

—. Profesionalni grupi [Trade Groups.] – In: Rodopi [Collection Rhodope Mountains], Sofia, BAS, 1994, p.361-362.

—. Sotsialni grupi I obshnosti. [Social Groups and Communities] – In: Strandja [Collection Strandja], Sofia, 1996, p.361-362.

—. Etnichni traditsii v lokalnite obshnosti [Ethnic Traditions in the Local Communities] – In: Izsledvaniya v chest na chl. kor. Prof. Strashimir Dimitrov [Studies in Commemoration of Corresponding Member of the Academy, Professor Strashimir Dimitrov] Studia Balkanika 23, p.763-764.

—. (ed.)Ekologia I traditsii. [Ecology and Traditions] – In: Kulturna ekologia [Cultural Ecology] (Collection). Sofia, 1997, p.327-320.

—. Esnafski organizatsii, [Professional Organizations] – In: Rodopi [Collection Rhodope Mountains], Sofia, 1996,

Boncheva, Tz. Brak i semeistvo pri katolitsite v Plovdivsko prez parvata polovina na XX vek [Marriage and Family Among the Catholics in the Region of Plovdiv During the First Half of the 20th Century]. Sofia, 2005.

Chukleva, M. Za antichniya proizhod na dva lozarski sada: chebar I korab. [About the Ancient Origin of Two Viticulture Vessels: Chebar and Korab] (Using materials from the Burgas District), *Izvestiya na bulgarskite muzei* [News from the Bulgarian Museums], Vol.1, Sofia, 1971, p. 212-216

Dennel, R. Early Farming in South Bulgaria from the 6th to the 3rd Millennium B.C., Oxford, 1978.

Drazheva, R. Trudovi obichai i obredi [Labour Rites and Rituals] – In: Etnografia na Bulgaria [Ethnography of Bulgaria], Vol.3, p.216-224.

Dukov, L. Zemedelieto I zheleznite zemedelski oradia prez Antichnostta v bulgarskite zemi [Agriculture and the Iron Agricultural Tools in the Bulgarian Lands in Antiquity] – In: IEIM, Vol.8, 1965, p.42-58.

—. Zhivotnovadstvo [Stockbreeding] – In: Dobrudja [Collection Dobrudja], Etnografiski, folklorni i ezikovi prouchvania [Ethnographic, Folklore and Linguistic Studies]. Sofia, BAS, 1974, p.230-249.

Ethnogenezis I etnokulturno nasledstvo na bulgarskia narod, [Ethno-Genesis and Ethno-Cultural Heritage of the Bulgarian People], Sofia, 1971, p. 87-88.

Etnografia na Bulgaria [Ethnography of Bulgaria}. Vol.1-3. Sofia, BAS, 1980, 1983, 1985.

Eyre, A., T.Wachtendorf, and G. R. Webb, Introduction."Bringing Culture Back In: Exploring the Cultural Dimensions of Disaster," – In: A.Eyre, T. Wachtendorf, and G. R. Webb (eds.). The Popular Culture of Disaster. – In: University of South California, School of Policy, Planning and Development, Issue Vol.18, No1, March 2000, p.5-19.

Filcheva V., Semeistvoto v Godechko [Family Structure in the Region of Godech], (Diploma Paper at the Ethnography Chair of the History Department of the St Kliment Ohridski University of Sofia), Sofia, 1984, p.38-39.

Genchev, S., Obichai I obredi za dajd [Customs and Rituals for Rain] – In: Dobrudja [Collection Dobrudja]. Sofia, BAS, 1974, p.346-351.

—. Narodna kultura I etnografia [Folk Culture and Ethnography], Sofia, 1984.

Georgieva, I. Etnografsko edinstvo na svatbata u bulgarite. [Ethnographic Unity of the Wedding Among Bulgarians.] – In: Etnogenezis I kulturno nasledstvo na bulgarskia narod [Ethnic Genesis and Cultural Heritage of the Bulgarian People]. Sofia, 1971.

Georgieva, I. Obichai pri svatba [Marriage Customs] – In: Pirinski kray. Etnografski, folklorni I ezikovi prouchvaniya [Pirin Region. Ethnographic, Folklore and Linguistic Studies]. Sofia, 1980.

Georgieva, I.(ed.). Armanite v Bulgaria [The Aromanians in Bulgaria]. Sofia, 1998.

Georgieva, Tz. Pomatsi – bulgari myusyulmani [Pomaks: Bulgarian Muslims.] – In: Obshnosti I identichnosti v Bulgaria [Communities and Identities in Bulgaria]. Sofia, 1998, p.286-308.

Grebenarova, S. Kalendarni obichai I obredi [Calendar Rites and Rituals] – In: Strandja [Collection Strandja]. Sofia, 1996, p.305-350.

Hristov, P. Predbrachno obshtuvane v selo Noevtsi, Breznishko prez 20-te – 30-te godini na XX vek [Premarital Contacts in the Village of Noevtsi, Breznik Region, in the 1920s-1930s], Bulgarska etnografia [Bulgarian Ethnography]. 1990, No6, p.9-10.

—. Sotsialna marginalizatsia I narodna kultura. [Social Marginalization and Folk Culture] – In: Etnografski etudyi. [Ethnographic Studies]. Sofia, 2006, p.39-46.

Kanev, K. Minaloto na selo Momchilovtsi [The Past of the Village of Momchilavtsi.]. Sofia, 1974.

Karamihova, M. Uzhasniyat "obraz" na turchina kato brachen partnyor [The Horrible "Image" of the Turk as a Marriage Partner] – In: Predstavata za drugia na Balkanite [The Conception of the Other in the Balkans]. Sofia. 1995, p.263-267.

Krastanova, K. Traditsii na trudova vzaimopomosht v bulgarskoto selo [Traditions of Mutual Assistance in Work in the Bulgarian Village]. Sofia, 1986.

Krasteva-Blagoeva,E., Lichnoto ime v bulgarskata traditsia [Personal Name in the Bulgarian Tradition]. Sofia, 1999.

Makaveeva, L. Bulgarskoto semeistvo: Etnosotsialni aspekti [The Bulgarian Family Ethno-Social Aspects], Sofia, 1991.

Manova, Ts. Pomeshteniya za novobrachni v Pernishko v kraya na XIX – nachaloto na XX vek [Premises for Newly Weds in the Region of Pernik at the End of the 19th – Early 20th Century], Bulgarska etnografia [Bulgarian Ethnography], No2, 1975, p.50-59.

Marinov, V. Za etnokulturnite prinosi na bulgarskoto zhivotnovadstvo [About the Ethno-Cultural Contribution of Bulgarian Stockbreeding] – In: Etnogenesis I etnokulturno nasledstvo na bulgarskiya narod [Ethno-Genesis and Ethno-Cultural Heritage of the Bulgarian People]. Sofia, 1971, p.87-88.

—. Ovtsevadstvo I kozevadstvo [Sheep-Breeding and Goat-Breeding] – Etnografia na Bulgaria [Ethnography of Bulgaria], Vol.2, Sofia, BAS, 1983, p.43-50.

Markova, M. Traditsionnata hrana I tipologiyata na zemedeltsi-skotovadtsi [Traditional Meals and Typology of Agrarians-Stock Breeders], Bulgarska etnologia [Bulgarian Ethnology], XXXII (2006), No2, p.161-173.

Mocheva, H. Selskoto zemedelso domakinstvo v Bulgaria prez 1935-1936 g. [Village Agrarian Household in Bulgaria in 1935-1936]. Sofia, 1938.

Odum, H.E. Energy Basis for Man and Nature, New York, 1976.

Peneva, L. Lozarstvo [Viticulture] – In: Etnografia na Bulgaria [Ethnography of Bulgaria], Vol.2, Sofia, 1983, p.26-30.

Pesheva, R. Semeistvoto I negovoto istorichesko razvitie [The Family and Its Historical Evolution] – In: Etnografia na Bulgaria [Ethnography of Bulgaria], Vol.1, Sofia, 1980, p.286-302.

Pimpireva, Zh. Karakachanite v Bulgaria. [Karakachans in Bulgaria.]. Sofia, 1998.

Popov, A., K.Pavlov. Rastenievadstvo [Plant-Growing], part 1, Sofia, 1966.

Primovski, A., Bit I kultura na rodopskite bulgari [Lifestyle and Culture of the Rhodope Bulgarians], SbNU, No54, Sofia, 1973.

Radeva, L. Traditsionna narodna hrana [Traditional Meals] – Etnografia na Bulgaria [Ethnography of Bulgaria], Vol.2, Sofia, BAS, 1983, p.291-295.

—. Hranene pri bulgarite [Nutrition of Bulgarians] – Etnografia na Bulgaria [Ethnography of Bulgaria], Vol.2, Sofia, 1983, p.297-298.

Raichevski, S. Zemedelie [Agriculture] – In: Strandja [Collection Strandja], Sofia, BAS, 1996, p.28-29.

Raichevski S. Bulgarite mohamedani [Bulgarian Muslims]. Sofia, 1998, 2nd edition 2004;

Rakshieva, S. Zhivotnovadstvo [Stockbreeding] – In: Loveshki Krai [Collection Lovech Region]. Sofia, BAS, 1999, p.31-60.

Sanders, I. Snosheniyata na edno bulgarsko selo s okolnata sreda [Relationships of a Bulgarian Village with the Environment] – Zemedelsko-stopanski vaprosi [Agricultural Matters], 1935-1936, No3, p.124.

Shishkov, S. Sledi ot kulta na oganya v Rodopite [Traces of the Cult of Fire in the Rhodopes] Rodopski napredak [Rhodope Advancement], 1909-1910, No5, p.5-11.

—. Zhivotnite v svetogleda na rodopskite poseleniya [Animals in the World Outlook of the Rhodope Settlements], Rodopski napredak [Rhodope Advancement], 1912, No4, 5, 7, 10.

—. Shishkov, S. Bulgaro-mohamedanite (pomatsi) Istoriko-zemepisen i nauchen pregled s obrazi, [Bulgarian Muslims (Pomaks), Historical and Local History Scientific Review with Images], Plovdiv, 1936.

Stopanska istoria na Bulgaria [Economic History of Bulgaria], 681-1981, Sofia, 1981, p.373-376.

Todorova, M. Balkan Family Structure and European Pattern: Demographic Development in Ottoman Bulgaria, Washington, 1993.

—. Balkanskoto semeistvo. Istoricheska demografia na osmanskoto obshtestvo prez Osmanskia period [Balkan Family Structure: Demographic History of the Ottoman Society During the Ottoman Period], Sofia, 2002.

Totev, A. Sastoyanieto na prenaselenost v nasheto zemedelsko stopanstvo [Overpopulation in Bulgarian Agriculture], Spisanie na ikonomichesko druzhestvo [Journal of Economic Society], XXXIX, 1940, No6, p.353-384.

Tzaneva, E. Edna starinna praktika pri osinovyavane na dete. [An Old-Time Practice in Child Adoption], Bulgarska ethnografia [Bulgarian Ethnography], 1990, No2, p.23-34.

—. Simvolikata na ritualite pri osinovyavane na dete [Symbolism of Rituals in Child Adoption] – In: Etnografski problemi na narodnata duhovna kultura [Ethnographic Problems of Folk Spiritual Culture], Vol. 2, Sofia, 1994, p.25-34.

—. Antropologichni teorii za osinovyavaneto [Anthropological Theories of Adoption], Bulgarska etnologia [Bulgarian Ethnology], 1996, No2, p.51-64.

—. Osinovyavane, bibleiski etos I zakon. [Adoption, Biblical Ethos and Law] Bulgarska etnologia [Bulgarian Ethnology], 2003, No2-3, p.30-42.

Vakarelski, H. Etnografia na Bulgaria [Ethnography of Bulgaria], Sofia, 1977, p.91-92.

— Zemedelie [Agriculture] – In: Etnografia na Bulgaria [Ethnography of Bulgaria], Vol. 2, Sofia, 1983, p.15.

—. Etnografia na Bulgaria [Ethnography of Bulgaria]. second ed. Sofia. Nauka i izkustvo, 1977.

Vassileva, M. Kalendarni praznitsi I obichai [Calendar Holidays and Rites and Rituals] – In: Dobrudja [Collection Dobrudja]. Etnografiski, folklorni i ezikovi prouchvania [Ethnographic, Folklore and Linguistic Studies]. Sofia, 1974, p.304-306.

—. Kalendarni praznitsi I obichai [Calendar Holidays and Rites and Rituals] - In: Etnografia na Bulgaria [Ethnography of Bulgaria], Vol.3, Sofia, BAS, 1985.

Zhelyazkova, A. Turtsi. [Turks.] – In: A.Krasteva (ed.). Obshnosti I identichnosti v Bulgaria. [Communities and Identities in Bulgaria]. Sofia, 1998, p.371-397.

Zlatarev, I. Ptitsevadstvo v Bulgaria [Poultry Raising in Bulgaria] Journal of the Bulgarian Economic Society, 1903, No6.

Illustrations

1. An Icon of St.George, the Saint who was believed to master the rain and fruitfulness

2. A Ritual Mask of a Kouker; Koukeri dances were performed by young unmarried men to ensure good harvest and protect it from natural disasters

3. Part of a ritual group "Surva" in Mid West Bulgaria; dances played by boys and young men at the beginning of the year with the purpose to scare away the evil forces

4. Ritual ploughing, part of the "Koukeri" - Ritual in Southeast Bulgaria

5. Female and Male Ritual Figures of "Martenitsi" – made and worn by each family from the 1st March, in order to protect the family members from diseases and misfortune

Abstract

This study of the Bulgarian traditional ecology is an attempt to outline the stable responses of the Bulgarian rural population to the difficult environmental, social and political conditions over the last centuries. Traditional mechanisms of natural and social adaptation and their expression in cultural forms and social interrelationships (family, household, social interaction, sacrifices, rites and holidays) are investigated to present the holistic response to environmental and social crises in a long-term perspective.

The empirical data collected and analyzed show that the Bulgarian traditional culture (at least in the last period of its existence, in 18th-20th centuries) responded to the challenges of nature through building a specific structural element on the border of rational and spiritual concepts named here "Bulgarian Traditional Ecology", which served in different historical periods as an adaptive strategy for coping with crises and disasters.

PhD Tanya Boneva is an Associated Professor in ethnology at the Department of Ethnology, Faculty of History, Sofia University St Kliment Ohridski. Her main fields of interest are social forms and institutions in Bulgarian society, traditional ecology, migrations and culture, power and identity. She delivers lectures in Balkan Ethnology, Traditional institutions on the Balkans and Europe, Culture and Identity.

IDEAS AND BELIEFS OF THE BULGARIANS ASSOCIATED WITH NATURAL DISASTERS

NIKOLAY KOLEV
ASSOC. PROFESSOR, DEPARTMENT OF ETHNOLOGY,
UNIVERSITY OF VELIKO TARNOVO

The underdeveloped productive forces in the tribal system, man's dependence on Nature and the uncontrollable elements, the limited experience and knowledge (or the lack of such), are among the objective prerequisites for the existence of a naïve and primitive world view.

An irrational world view is observable invariably where an explanation is sought for the nature of the natural forces and elements, they being spiritualized and, later on, deified, as well.

Myths were the earliest reflections of human intelligence on the causal relationships between the phenomena in Nature. They were also the earliest, though naïve, elementary attempt at explaining, studying and mastering the forces of nature and the natural elements. Therefore, quite a number of scholars define myths as primitive science, as "primordial natural philosophy". Primitive man shaped the world with the help of his imagination and believed in mythical beings and elements/forces (wood nymphs, dragons, lamias, whirlwinds), which personified the natural elements. The feeling of fear was intertwined with veneration and elements of cult of such beings. Gradually man started to deify the forces of nature and the natural elements.

The medieval heathen world view of the Bulgarians was re-interpreted in a Christian spirit and acquired a Christian form. However, the pagan views and rituals did not die out straightaway; they continued to exist. So, as a result a mixture was created out of the two ideologies – the heathen and the Christian one – and the so-called folk Christianity took shape, whereby some of the old pagan ideas and holidays were adapted, tied up with the formal Christian calendar and intertwined with the cult of the Christian saints.

While the prevailing attitude of man towards the environment and the forces of nature was manifest in the pagan conceptions and practices which aimed at influencing the latter and modifying them to the benefit of man, in the case of the Christian cult, what was at hand was a worship of saints who were believed to secure fertility and prosperity. This worship found expression in the construction of chapels and churches, saying of prayers and observance of bans, as well as celebration of feast days in honor of the saints-patrons of the forces of nature and of the natural elements.

The natural calamities occurring without the interference of man are those associated with some *cosmic* events (such as solar outbursts, falling comets, solar and moon eclipses), *atmospheric* ones (hailstorms, lightning and thunder, drought, sudden torrential rains, abundant snowfall, tempestuous winds of varying force – whirlwinds, tornados, typhoons), and *earthly* (earthquakes, volcanoes, avalanches) events.

The appearance of a comet or the falling of a meteor was known among the Bulgarians as the appearance of ***a portentous star, a grand star,*** **a** ***grand star with a tail, a kuirukliya [tail] star, a light star***. These events were interpreted as the appearances of a fiery dragons flying like twisted beams over the village, and was considered to be a sign of a pending calamity. These extraordinary phenomena spelt a forthcoming disaster: a disease among the farm stock and the people, wars. Seeing a great number of stars fall, or ***tailed stars,*** Bulgarians used to predict the death of a great number of people as a result of an epidemic, starvation or war.

As rare phenomena they were recorded in numerous notes and chronicles: "It should be known that ***a portentous star*** appeared in the sky on the 13th day of August and stayed until November. This happened in the year 1812 AD. And at that time there was a severe winter, the winter stayed for very long. And we found it very hard to find food; there was great suffering and cold that winter. And that star appeared then to warn us, brothers, and there was much disease at that time."[1] As a coincidence this happened to be the year of Napoleon's invasion in Russia.

Solar and moon eclipses[2] also struck the imagination of the Bulgarians as unusual occurrences. The moon eclipse was most often explained by the act of witches stealing it from the sky, transforming it into a cow and milking it[3]. The solar and moon eclipses caused fear among the population, because they were believed to portend the coming of bad days and dis-

---

[1] Pisahme da se znae. Pripiski I letopisi [We Wrote This Down to Be Known. Notes and Chronicles]. Sofia, 1984, See: notes on p.80, 140,142, 159, 188, 289, 301, 320, 322, 339.

[2] Ibid, p.84, 303, 305.

[3] H.Vakarelski. Etnografia na Bulgaria [Bulgarian Ethnography]. Sofia, 1974, p.487.

eases, hunger, death and other ills, which the Lord was about to send for great sins committed[4].

The solar and moon eclipses were feared by the people as auguring a forthcoming misfortune and ills: a disease, war, starvation and death. The eclipse was believed to be the result of magic[5]. Widespread in Northwest Bulgaria was the belief that the solar and moon eclipses were caused by ***varkolaks [ghoul]*** or dragons and lamias, which were said to have swallowed the light of heaven.

Widespread was the myth about the solar eclipse as caused by animals and monsters. The conception that snakes drink the sun is often found among the Slav peoples – Bulgarians, Poles, Byelorussians, and Ukrainians[6]. An explanation very close to the scientific one about the solar eclipse, caused by the moon, is found in John the Exarch's writing *Nebessa [Heavens]*: "And the sun is darkened when the body of the moon stands as an obstacle, a shade which does not allow it to shed its light to us".[7]

The concepts the Bulgarians developed of the atmospheric phenomena were closely connected with their work. Various practices are known whose purpose it was to allow man to exert an influence over the atmospheric natural elements – an expression of the human desire to control and subject them. The power of observation entails the establishment of actual connections and laws, while an irrational element is manifest when an explanation is sought about their real nature.

The traditional folk beliefs had it that the clouds were locked in caves and kept by some lamia or, in another version, by a blind old man. Their master, however, was though to be St Elijah, also called by the people ***Thunder Saint.*** According to other beliefs, this was our Lord Himself, whereas, depending on the season, they were controlled by different saints: in the springtime the clouds were thought to be controlled by St George, in summer – by St Elijah, in autumn – by St Demetrius, and in winter – by St Athanasius (Illustration 1).

In the region of Plovdiv the clouds were considered to be denser mists which had risen high into the sky. When they were not seen, people said that they "have gone to drink their full of water" from the sea and the riv-

---

[4] D.Marinov. Narodna vyara I religiozni narodni obichai [Folk Beliefs and Religious Folk Rites and Rituals]. – SbNU, Vol.28, 1914, p.34.

[5] Ibid, p.141.

[6] I.Georgieva. Bulgarska narodna mitologia [Bulgarian Folk Mythology], Sofia, 1982, p.487.

[7] P.Petrov, V.Gyuzelev. Hristomatia po istoria na Bulgaria [Reading Book on Bulgarian History], Sofia, Vol.1, 1978, p.321.

ers, or that the wind had ***pushed them [namushil] into*** the ravines. Dark, low clouds were known to bring rain, whereas thick, light-coloured clouds with dark ***clusters [valma]*** at the fringes, which ***roar [fuchat]*** while moving, bring hailstorm[8].

Dark clouds were looked at with fear and anxiety, because almost invariably coming with them are great floods, devastation and hailstorms.

According to D. Marinov, the dark or black cloud is represented in folk beliefs as ***a hala [whirlwind]***, which flies with the wind and causes devastation wherever it passes. Such a black, dark cloud is believed to consist of sticks and these sticks are the wings and the tail of the *hala*. Fighting against such a *hala*-cloud are dragons and eagles. If in a village courtyard there is a ***dragon*** or ***a cross eagle***, they fly high up, fight the *hala* and drive it elsewhere[9].

It was believed that no hail falls on a field boundary where there is an eagle. For that reason eagles were particularly respected, and in Sofia region a *kurban* offering – a lamb shoulder[10] – was made to them.

The concept of the *hala [whirlwind]* as a mythical being, carrying devastating rain, hailstorm and winds, was especially characteristic of Western Bulgaria. In the rest of the country it is encountered in folklore as a poetic metaphor of rain, whereas in the folk speech the word is also used to mean strong wind, and whirlwind.

The belief in the economic function of the dragon was characteristic of Western Bulgaria again. Here almost every village used to have a personal dragon-protector, which guarded the lands from lamias, whirlwinds, and other village dragons, as well as from hailstorms and other calamities. Dragons would climb down into the villages from the caves they inhabit in order to guard the fields. They would perch on age-old oak trees, which were their own, wherefrom they would protect the field boundaries and the fields from hailstorms.

There are three such oak trees known to the population of the village of Lokorsko, Sofia region. On the feast day of the Holy Trinity, when ointment was put on them, the villagers came with icons and a clapper headed by the priest; they made a round of the fields and stopped in front of these trees, because it was believed that they guarded the village from hail-

---

[8] Zh.Stamenova. Naroden mirogled [Folk World View]. – In: Plovdivski krai [Collection Plovdiv Region]. Sofia, 1986, p.296.

[9] D.Marinov. Narodna vyara I religiozni narodni obichai [Folk Beliefs and Religious Folk Rites and Rituals] - In: Izbrani proizvedenia [Selected Works], Vol.1. Sofia, 1994, p.62.

[10] I.Georgieva. Mirogled na bulgarskia narod [World View of the Bulgarian People]. – In: Etnografia na Bulgaria [Ethnography of Bulgaria], Vol.3, 1985, p.45.

storms, floods or torrential rains. When a dark cloud came over the village, winged dragons flew out of the cloud and descended on these oak trees to fight against the whirlwinds until they chased them away. Due to the power of the ointment the dragons conquered the whirlwinds and drove them away. No one dares to cut these holy trees fearing lest he should be harmed[11].

According to traditional folk conceptions thunder and lightning were caused by different agents, but most often by the two-wheeled chariot of St Elijah. As soon as the saint noticed some storm threatening to ruin the crops of the fields, he called the dragons and made them climb onto his chariot, and, together with them, he aimed his spear at the storm to shoot it through. Lightning and thunder were produced when he thrusted his spear at the storm and drove his chariot over the clouds, whereas the storm, in order to protect itself, sprinkled the water it kept in its mouth onto the dragons: drops of that water fell onto the earth in the form of rain[12]; the water was believed to come either from the barrels which our Lord, or St Elijah, rolled, or from the arrows which the Lord shot at the dragon, or from St George (Illustration 2) and St Elijah, who walked in the sky and shot. According to folk beliefs, St Elijah controlled thunder and lightning.

According to another set of beliefs, thunder and lighting resulted from the battle of dragons. It was believed that the dragon appeared as "a fiery warped beam in the sky, whereas the lighting was the dragon's arrow, with which he fought the whirlwind." It thundered when two dragons fought, and when a dragon opened its mouth there was lightning. Other Bulgarian legends have it that the dragon itself turned into a fiery arrow, and in the struggle with the whirlwind St Elijah sent three little arrows – little dragons, which were his fiery arrows. According to I.Georgieva, this is the fiery nature of the Bulgarian dragon.

Lightning was believed to be the dragon's arrow with which it drove the whirlwind away; after being sent, the arrow hid in a tree, in a farm animal or a man, and if it got into a farm animal or a man, the latter died. That is why pitchers and other vessels for water should not be left without a lid in the open, because the arrow might hide in them.

Later on, this conception was transformed to conform to the Christian world view. It came to be believed that thunder was our Lord's arrow with which God chased the Devil. The Devil hid in a man, and the man was in

[11] Ibid, p.82; see also Tz.Ginchiev. *Trud* [Labour], II, 1880, p. 60.
[12] I.Georgieva. Op. cit., 1985, p.83.

danger of being killed as a "partner of the devil"[13]. Here the function of "a dragon-fighter" was assigned to the Lord (Illustration 3&4).

There was a common belief that on the place the arrow had fallen a white clean stone was formed, which might be rounded or with an iron edge. It was called ***'a stone on which a thunder has fallen' [grom padina, treshten kama], or 'God's arrow' [Bozha strela].*** This stone used to be carried to bring you health and to avert thunder. It was believed that it also cured ***epilepsy [vonkashka illness]*** and the sick were given water to drink in which the 'arrow' had been immersed[14].

In the Middle Ages, books of prophecies were widespread with predictions of what health and fertility would be during the year, judging by the first thunder: should it come in January, it was considered to be an evil omen for great people; in February – for humble people; the closer the date thunder was heard to the feast day of St George the better it would be for the crops; or as many days as were left to the feast day of St George, so much the crops would cost; thunder from the North meant that the harvest would be poor; thunder from the West augured well for the Bulgarians; if it thundered from the East it foretold a good harvest; if it thundered before noon, it meant that there would be fertility; if it was in the afternoon, it was not a good omen; if a crack of thunder was heard early in the year, what was expected was starvation and expensive grain[15].

Magic actions were performed at hearing the first thunder of the year which involved stimulation or forewarning. Three little pebbles were cast up with the words, "Knock, knock big field." The baskets, casks, and the barn were hit with a stone, a stick or a washing batter, so that grapes and wheat might abound. People knocked their heads with a small stone or a piece of iron, saying, "Knock, knock stone (iron) head", so that they might not be struck by thunder and might not suffer from headache"[16].

Back in the Middle Ages there were also attempts at rational explanations of the origin of thunder and lightning. According to one such explanation, these resulted from the clouds rubbing to one another, while another had it that these were produced by the wind which was kept within the clouds, in their voids, and got free forcefully. What was observed that during winter there were no thunders was explained by the fact that in that season the thickness and moisture of the clouds were greater. As far as

---

[13] Ibid, p. 84; M.Arnaudov. Ochertsi po bugarski folklor [Sketches of Bulgarian Folklore], Vol.2, 1969, p.602.

[14] I.Georgieva. Naroden Mirogled [Folk World View.] – In: Pirinski krai [Collection Pirin Region]. Sofia, 1980, p.463.

[15] I. Georgieva. Op. cit., 1985, p.40.

[16] D.Marinov. Op. cit., 1871, p.65; Zh. Stamenova. Op.cit., p.297.

lightning is concerned, it preceded thunder because the sound could not be heard immediately.

Hailstorm is a kind of precipitation which consists of roughly spherical pellets (balls or pieces of ice) of different size – from the size of a pea to that of an egg, the diameter of the hailstones ranging from a few millimeters to 5 and more centimeters. For instance, on May 21, 1977 hailstorm raged in the region of Pernik (Elov Dol, Mureno and Padine), the weight of a single icicle being 0.8 kg.

Hailstorms form in powerful thunderstorm clouds when drops of cool water collide with ice crystals. They fall mainly during the warm part of the year: from April till October. Parts of the regions of Montana, Vratsa, the Ludogorie, the Sredna Gora Mountain, Panagyurishte, Blagoevgrad, Chirpan and the Eastern Rhodopes are most often and most strongly affected by hailstorms. Hailstorms are often accompanied by torrential rains, which increase the damage to agriculture, destroying the crops, hurting and killing people and farm animals in the fields.

This atmospheric phenomenon has left lasting traces in the life of the Bulgarian people – a fact evidenced by its reflection in notes and chronicles, as well as in folk beliefs and practices developed to protect against hailstorms.

"August 15th, 1959. ***A strong hailstorm raged*** and it flooded the village of Arnbanasi"; on September 14th, 1846, Saturday, ***there was a strong hailstorm;*** during the night there was a strong tempest. ***The hailstones were as large as eggs.*** There were no victims. The hailstorm went on for an hour and the rain continued for another hour, and the ice cover remained until the following day"; "The year 1847, the 15th day of August,. ***There was a strong hailstorm which broke tiles and killed numerous livestock";*** "June 25th, 1856, ***at 11 a.m. there was a strong hailstorm,*** the hailstones weighing 25 grams each"; "***It hailed, the hailstones as large as eggs. Quite a number of tiles were broken*** on houses and other buildings. It went on for half an hour and ***struck all the vineyards and devastated the fields.*** None of the old people remembered having witnessed such large hails, May 3rd, 1871."; "On August 25, 1860, on Saturday evening, around 10 o'clock, ***it hailed so much that it broke the tiles of the houses and struck the vineyards so heavily that their twigs were broken, and a great number of farm animals were killed and wounded in the fields. And it crushed the bulrush to the ground".*** "The 25th day December, 1860, ***a loud thunder was heard and it hailed"***[17].

[17] "We Wrote So That It Could Be Known", Op. cit., p.102, 130, 162, 185, 322, 351.

Two conceptions, two explanations, prevailed among the Bulgarian people: a rational and an irrational one. The rational explanation is that the hail formed out of dense water in the clouds, out of the raindrops, dew and vapour, due to the cold wind or the cold.

According to the irrational conceptions, the hail was caused by the elements, the devil, by three saints, namely, Bartholomew *(Vartolomei)* [The Whirlwind Breaker], German and Marc the Hailstorm *(Marko Gradobitnya)*, or by three men – The Crooked, the Blind, and the Deaf (*Krivcho, Slepcho* and *Gluhcho*). That last concept was widespread in the region of Kyustendil.

It was also believed that the hailstones were made by the sinners (drowned or hanged people, prematurely born babies and such that died before being baptized, as well as babies born out of wedlock) in the Netherworld. According to other beliefs they were produced by the wood nymphs which accumulated hailstones in the lakes and then lifted them into the clouds; or else the hailstorms and the clouds which released hailstones were believed to be dragons[18].

The most common traditional Bulgarian concept of the hailstorm is that this natural calamity was the result of a sin committed in the village, usually associated with breaking the norms of public moral and social behavior – i.e. thefts, rape of a maiden, murder of a baby born by a maiden or a widow in adultery, removal of the nest of an eagle, non-observance of the customs through which hail is prevented.

It was also believed that a hail-cloud was preceded by a fledgling called ***'orle' [fledgling eagle], sukle, kyoravo pile [blind fledgling bird], navyak***, which was most often conceived as blind and leading the cloud in no particular direction. In the region of Plovdiv, flying in front of such a cloud was thought to be a ***crastat orel [cross marked eagle]***, which "screamed and fought" the hail-cloud in order to scatter it. In considerably rare cases people believed that the eagle directed the cloud and indicated where it was to drop the hailstones[19].

A whole range of magic ritual actions were developed in traditional folk practice to prevent or stop the hail, to turn back or divert the hail-bearing cloud.

Men would shoot their guns to the hail-cloud to scatter it or kill the eagle which leaded it. With the same purpose they would ring the church bells or beat on iron sheets.

---

[18] I.Georgieva, Op. cit., 1985, p.40; Bulgarska mitologia, Entsiklopedichen rechnik [Bulgarian Mythology, Encyclopedic Dictionary], 1994, p.149.

[19] Zh.Stamenova, Op. cit., p.297.

In "the dead of night" two twin brothers would start ploughing the common village pasture with two twin oxen; they would bury the first Easter egg in the field, leaving offerings on the field boundaries and speaking incantations; hence the belief that "the hail falls up to a border, to a field boundary"[20]. Featuring in the incantations and evil spells are the fear from and a plea to the incoming disaster, "Upwards, Kuzman, Damyan, upwards, where no maidens sing, no children cry, where no cock sings, where no ax would cut!" or, "Go back little cloud, Kuzman, go back, German, it is a great holiday for us and there is no room for you. Go to the deserted forest, where no cock would sing and no lamb would bleat!" Unable to control the natural elements man resorted to actions which had magic connotations and were of ancient pagan origin, actions which, in the final count, did not alleviate in any way the disastrous situation of the peasant. The personal name German was known as early as in the 6th century A.D., under the Emperors Justinian and Mauricius, and was associated with the Slavonic word *gram [thunder]* – that's where the origin of the pseudo-scientific myth upheld by P.R.Slaveikov could be sought – to the effect that for the Old Bulgarians German or Garmene was "the god of thunder and the city", transformed, later on, into the Christian saint German.

Working on the assuming that old mythology did not know of the existence of such deity, regardless of Slaveikov's interpretations, the fact is that, indeed, the word 'gram' [thunder] might well have sounded like the personal name German, and since St German's feast day in the religious calendar fell during the period of the summertime tempests and hazardous hailstorms, it was easy for a legend to develop and survive with all its practical consequences and incantations. The population of Northwest Bulgaria, in particular, observed this feast day with great respect in view of the fields being under considerable threat of hail; and in the course of the celebration pleas and threats were addressed to St German, transformed into a kind of demonic being German, asking him to have mercy on the people, crops and livestock and drive the natural disaster to some other place (Illustration 5).

Featuring among the irrational practices for averting hail are also: the lighting of the "Varvara"; the rituals performed on "St George's feast day", the "Easter" or "Christmas Eve" candle; the casting of a red-dyed Easter egg against a hail-bearing cloud; the kindling of the willow twigs sanctified in church or looking at the clouds through them; the turning upside down of the over-the-fire stand on which pots are placed; putting an

[20] I.Georgieva, Op. cit., 1980, p.463.

axe, a hoe, a scythe or a knife with the sharp blade upwards, so that the hailstones could be "cut short", as well as the hail itself; taking out in the open the ***gribach*** [the poker used to pull live coals out of the hearth] and the ***pometa*** [a piece of cloth used to clean the oven after it had been used]; taking out the round piece of wood on which women knead the dough along with the rolling-pin; making the eldest or the youngest child in a family to swallow one or three hailstones; casting of salt against the hail-bearing cloud. The ***Kapantzi*** community practiced taking out and placing some salt and pepper next to iron objects – the salt melts the hailstones, while the iron objects (a stand on which pots are placed over the fire, an ax, a scythe, a knife) play an apotropeic role. Used was processed iron, which had passed through fire, through which process to the power of iron was added the power of fire, repulsing and driving away the evil forces[21].

What was practiced in the region of Kyustendil was a naked woman driving back the hail-cloud by means of incantations, a red-dyed Easter egg Easter or a candle preserved from Christmas Eve[22].

It was commonly believed that no hail falls on the fields of people who give money or objects to charity. D.Marinov quotes such an occasion, "There was a peasant who would always give something to every poor man (or orphan, or widow); he would never turn anyone back empty-handed. Once, in the summertime, when a dark cloud appeared, a voice was heard speaking from the cloud, "Whirl, whirl, Bartholomew, strike, beat, German, but spare Kutlyova field!" And the hail fell strongly and devastated all the fields, but when it reached the field of that peasant – the Kutlyova field – it passed round it and did not do any damage to it"[23]

Featuring in the Bulgarian calendar of holidays are more than 18 days during which rituals are performed and bans are observed aimed at ensuring protection from that natural calamity – hail.

In Western Bulgaria, on Christmas, the host addresses an incantation to German, the patron saint of hails, which has a protective, apotropeic function, "German, German, little cloud! Come to have supper with us. May you come now, and in the summertime may we not see your eyes either in the field or in a meadow. Now we have prepared bread, stewed dried fruit,

---

[21] D.Todorov. Naroden mirogled.[Folk World View.] – In: Kapantsi [Kapantsi Collection]. Sofia, 1985, p. 263; D.Marinov. Op. cit., 1994, p.67.

[22] P.Ts. Lyubenov. Sbornik ot razni umotvoreniya iz kyustendilsko. [A Collection of Diverse Folk Tales from Kyustendil Region.], 1896, p.68.

[23] D.Marinov. Op. cit., 1914, p.30.

beans, nuts, seeds, white onions, everything is full, and most of all brandy."[24]

Lighting fires on Cheese-fast Sunday was also connected with driving hail away. The fires were built up as big as possible, because it was believed that no hail would strike within the perimeter lit by flames of the fire[25]. In the village of Trivoditsi, Plovdiv region, in the evening on Cheese-fast Sunday people traditionally shoot in the chimney and throw some water upwards in it so that no hail may fall[26].

In the region of Plovdiv, a red piece of cloth or a woman's kerchief was washed on that Tuesday referred to as ***dry Tuesday,*** or ***black Tuesday***, people sinking their fingers into the soapy water; ritual bread was baked into an oiled pan so that there might be no hail[27].

The rites and rituals associated with March 1st were also determined by a number of prophylactic, protective and stimulating measures associated with the process of the cyclical renewal of nature in the transition from winter to summer. There were ***taboos*** (bans) to working, eating some foods, wearing some colors clothes, etc. (Illustration 6).

So that no thunder may fall, in some villages in Dobrudja plain women did not weave on the first three days of March; they did not beat rugs (the noise of beating rugs resembles that of thunder); they did not boil anything, so that a cloud might not be "boiled"[28].

The following requirements are known to have been connected with the ban on colours (white and red): so that there might not be hails and frost during the year, nothing white was taken out on March 1st. On this

---

[24] Y.Zahariev. The Kyustendil Region – SbNU [Collection of Traditional Culture and Folklore], Vol.32, Sofia, 1918, p.163-164; A.Marinov. Narodopisni materiali ot Graovo [Folklore Materials from Graovo]. – SbNU, Vol.49, 1958, p.722; B.Stoilkov. Iz narodnata kultura na Busintsi, Transko [From the Folk Culture of Busintsi, Tran Region]. – INEM [Studies by the National Ethnographic Museum], Vol.12, 1936, p.144.

[25] M.Vassileva. Kalendarni praznitsi I obichai [Calendar Holidays and Rites and Rituals.] – In: Etnografia na Bulgaria [Ethnography of Bulgaria], Vol.3, Sofia, 1985, p.110-111.

[26] Zh.Stamenova. Op. cit., p.259.

[27] Ibid, p.262.

[28] M.Vassileva. Kalendarni praznitsi I obichai. [Calendar Holidays and Rites and Rituals.] – In: Dobrudja [Collection Dobroudja], Sofia, 1974, p.338; D.Sabotinova, Otkoleshno vreme, Kalendarni praznitsi I obichai ot Silistrensko [Time Past, Calendar Holidays and Rites and Rituals from the Region of Silistra], 1998, p.90.

day no washing was done and no white sheets were spread to dry in the open[29].

Magic force was assigned to the first Easter egg: it was believed to have the power to avert hail and heal. That is the reason why it was buried in the field on St George's feast day or on the first day of the sowing[30]. Still in warding off hail, the Thursday of the Easter All Souls' Day, as well as the following six Thursdays, is called ***Maundy Thursday.*** On these days women did not do any household work: they did not wash, did not beat rugs, did not spin, sow or weave, while men did not plough[31]. These bans were observed on each of the 12 Fridays called ***Good Friday***, so that again no hail would fall.

In some villages in the regions of Kazanluk and Gabrovo, practiced on the second day of Easter, called ***razturi, razmetni Monday***, was an exchange of eggs, the purpose of which was again to ensure protection against hail. Young bachelors and maidens got together in the green in the centre of the village and split into two groups: one group was of maidens and the other was of young men. The young men rolled dyed Easter eggs to the maidens; the latter received them, but sent them back to the young men[32].

It was further believed that on St George's feast day, when maidens sang over bunches of flowers, as far as the maidens' voices reach, there would be no hail[33]. On St George's feast day again, on which occasion they used to slaughter a ritual lamb, the people of the village of Suhodol, Sofia region, collected the blood of the lamb in a big earthenware pot, covered it with a lid and buried it in the earth. The blood in the ***kyup*** (the earthenware pot) played the role of a kind of meteorological station during the year. As soon as a cloud appeared in the sky, the old woman who kept the pot would go there, open the lid and look in: if the blood had started to boil, this was a sign that a hail-bearing cloud was coming; if it was not boiling, the cloud was just bearing rain[34].

---

[29] L.Mikov. Parvomartenska obrednost [March 1 Rites and Rituals]л Sofia, 1985, p.74-75.

[30] M.Vassileva. Op. cit., 1985, p.120.

[31] Ibid., p.120-121; R.Drazheva. Kalendarni praznitsi I obichai [Calendar Holidays and Rites and Rituals]. – In: Collection Pirinski krai [Pirin Region]л Sofia, 1980, p.447.

[32] M.Vassileva. Op. cit., 1985, p.121.

[33] R.Drazheva. Op. cit., p.447.

[34] D.Marinov. Op. cit., 1914, p.453.

It was on St George's feast day again that the straw from the Christmas Eve table was spread over the fields or put on fire again to ward off hail[35].

St Jeremijah (May 1st) was honored as a protector from thunder and hail, whereby the same bans quoted in connection with the other holidays were observed. It was believed that "St Jeremijah carried hail in his belt and should anyone not honor him and work on this day, the saint dropped the hail onto the earth". For this reason, nowhere was fieldwork done on this day, while women did not spin, weave, or turn the spinning wheel, lest they could attract thunder[36].

Listed among the holidays honoring hail is also the feast day of St German (May 12th). It was widely held that the saint was the worst in this group and, therefore, mostly in Western Bulgaria, he was called German, the hailstorm saint. If a hail-carrying cloud appeared in the sky on this day, candles were lit on the thresholds of the houses, on the outer side, and the cloud was chased away with the words pronounced out loud: "Carry it German, carry it in the faraway forests, where no grain is born"[37].

Widespread was the celebration of the feast day of Elena and Kostadin (Helena and Constantine) (May 21st) to ward off hail. It was connected with the belief that Elena carried hail in her sleeve or that Elena and Kostadin carried hail in a bag. In the Rhodope region, the following saying is common concerning the hail, "Sirnitsa makes it, Todoritsa throws it, while St Helena carries it in her sleeve"[38]. Some of the fire dancers performing on live coals on this day predicted hailstorms, pending diseases and war, as well as the ways to avert them[39].

Besides German, the Bulgarians honored three other hailstorm saints, namely, Vartolomei [Bartholomew] (June 11th), Elisei (June 14th) and Vidovden (June 15th). Being coined purely through folk etymology, their names were derived from some conceptions and practices associated with them. The first saint was celebrated on the feast day of the apostles Bartholomew and Barbara, and is also known as ***Vartomolyai [Turn and Pray], Vartolom [Turn and Break], Varti-udri [Turn and Strike], Vartilomi [Turn and Break]***. In Pirot the Thursday of the Wood Nymph Week is called Vartolom[40].

---

[35] M.Vassileva. Op. cit., 1985, p.122.
[36] R.Drazheva. Op. cit., p.338; Zh.Stamenova. Op. cit., p.273.
[37] A.Martinov. Op. cit., p.739.
[38] N.Haitov. Minaloto na s.Yavorovo [The Past of the Village of Yavorovo], 1958, p.150.
[39] N.Vassileva. Op.cit., 1985, p.125.
[40] I.Shishmanov. Prinos kam bulgarskata narodna etimologia [Contribution to Bulgarian Folk Etymology]. – SbNU, Vol.9, 1893, p.557.

In Struga, Kukush and Ohrid, as well as at some other places, people dis not plough and did not harness any draft animals, because they believe that on that day the sun turned from summer to winter and the ban would prevent whirlwinds.

The feast days of St Peter (June 29th) and St Paul (June 30th) were also observed, which was believed to protect from thunder and hail. In Eastern Bulgaria, ***Panaya*** and ***Tutusya*** are celebrated against hail on the first and second Mondays following the feast day of St Paul. During the so called 'dog days' (July 15th-17th) no field work used to be done, as well as no other work at all; the furnaces were not lit and no bread was baked; people did not wash, did not do any laundry, because the failure to observe these taboos was believed to entail fires and hails[41]. St Elijah, as patron saint of thunder (July 20th) was also honored in order to ensure his protection against hail. On his day a ***kurban*** offering was made – a cock or a hen was slaughtered. Very often, north of the Maritsa River, the offering was an ***obruk*** (a sacrificial offering), whereas in the region of the Pirin Mountain and south of the Maritsa River the offering was common to the whole village, sometimes related to a church holiday or a village gathering and was accompanied by a religious service calling for rain, against hail and thunder. A sheep or a ram, at some places even a calf, was slaughtered. These sacrificial offerings used to take place at a special site outside the village: on a mound, next to a chapel or a small church[42].

There were celebrations to ward off thunder and hail also on the feast day of St Mary (July 22nd). A home-made candle, called Varvara candle, was lit on the table on the feast day of St Barbara (December 4th); besides on this day, such a candle was also lit in the case of a difficult birth of a farm stock animal and in cases of thunder and hail.

Helpless in the face of natural disasters such as hail, unfamiliar with the natural laws like the other ancient peoples, the Bulgarian people also built up irrational concepts of that atmospheric phenomenon, resorting to magical methods to drive away the hail-bearing clouds; they honored saints considered to be patron saints in control of that natural element.

The beliefs, ideas and concepts, along with the cult for the saints associated with hailstorms, persisted in the consciousness of the Bulgarian people until the 1930s and 1940s, but had faded out of practice much earlier; a fact connected with the advent of scientific knowledge which radically modified the lifestyle of the Bulgarian people.

---

[41] M.Vassileva. Op. cit., 1985, p.130.
[42] Zh.Stamenova. Op. cit., p.276.

References

Arnaudov, M. Ochertsi po balgarski folklor [Sketches of Bulgarian Folklore], Vol.2, 1969.

Balgarska mitologia. Entsiklopedichen rechnik [Bulgarian Mythology, Encyclopedic Dictionary]. Sofia, 1994.

Drazheva, R. Kalendarni praznitsi I obichai [Calendar Holidays and Rites and Rituals]. – In: Pirinski krai [Pirin Region]. Sofia, 1980.

Georgieva, I. Naroden Mirogled [Folk World View.] – In: Pirinski krai [Collection Pirin Region]. Sofia, 1980.

—. Bulgarska narodna mitologia [Bulgarian Folk Mythology]. Sofia, 1982.

—. Mirogled na bulgarskia narod [World View of the Bulgarian People]. – In: Etnografia na Bulgaria [Bulgarian Ethnography], Vol.3, 1985.

Ginchiev, Tz. *Trud* [Labour-newspaper], II, 1880.

Haitov, N. Minaloto na s.Yavorovo [The Past of the Village of Yavorovo], 1958.

Lyubenov, P. Sbornik ot razni umoitvoreniya iz kyustendilsko. [A Collection of Diverse Folk Tales from the Kyustendil Region.], 1896.

Marinov, A. Narodopisni materiali ot Graovo [Folklore Materials from Graovo]. – SbNU, Vol.49, 1958.

Marinov, D. Narodna vyara I religiozni narodni obichai [Folk Beliefs and Religious Folk Rites and Rituals]. – SbNU, Vol.28, 1914.

—. Narodna vyara I religiozni narodni obichai [Folk Beliefs and Religious Folk Rites and Rituals]. In. Izbrani proizvedenia [Selected Works]. Vol.1, Sofia, 1994.

Mikov, L. Parvomartenska obrednost [March 1 Rites and Rituals], Sofia, 1985.

Petrov, P., V.Gyuzelev. Hristomatia po istoria na Bulgaria [Reading Book on Bulgarian History]. Sofia, Vol.1, 1978.

Pisahme da se znae. Pripiski I letopisi [We Wrote This Down to Be Known. Notes and Chronicles]. Sofia, 1984.

Sabotinova, D. Otkoleshno vreme, Kalendarni praznitsi I obichai ot Silistrensko [Time Past, Calendar Holidays and Rites and Rituals from the Region of Silistra], 1998.

Shishmanov, I. Prinos kam bulgarskata narodna etimologia [Contribution to Bulgarian Folk Etymology]. – SbNU, Vol.9, 1893.

Stamenova, Zh. Naroden mirogled [Folk World Wiew]. – In: Plovdivski krai [Collection Plovdiv Region]. Sofia, 1986.

Stoilkov,B. Iz narodnata kultura na Busintsi, Transko [From the Folk Culture of Busintsi, Tran Region]. – INEM, Vol.12, 1936.

Todorov, D. Naroden mirogled.[Folk World View.] – In: Kapantsi [Collection Kapantsi]. Sofia, 1985.
Vakarelski, H. Etnografia na Bulgaria [Bulgarian Ethnography]. First ed., Sofia, 1974.
Vassileva M. Kalendarni praznitsi I obichai [Calendar Holidays and Rites and Rituals.] – In: Etnografia na Bulgaria [Ethnography of Bulgaria], Vol.3, Sofia, 1985.
—. Kalendarni praznitsi I obichai. [Calendar Holidays and Rites and Rituals.] – In: Dobrudja [Collection Dobrudja], Etnografiski, folklorni i ezikovi prouchvania [Ethnographic, Folklore and Linguistic Studies]. Sofia, BAS, 1974.
Zahariev Y. The Kyustendil Region – SbNU, Vol.32, Sofia, 1918.

Illustrations

1. An Image of St. George on a Earring

2. St. George kills the *hala* – An Image on a Flask

3. Group of Ritual Figures – "Koukeri"

4. A Mask of a "Kouker"

5. "German"

6. "Martenitsi" hung on a tree

Abstract

The article presents the results of research on the traditional Bulgarian views, beliefs, and practical actions associated with critical situations, mostly natural cataclysms and epidemic deceases. Like all traditional rites of passage and rituals marking the transition from one life status to another associated with the calendar and the family customs, these offer rich and significant data for studying the mentality and the psychological basis of the Bulgarian traditional culture. In their efforts to fight the natural calamities the Bulgarians developed and performed various magical practices – many kinds of spells, apotropeic and contact magic, etc. – which are studied in this paper. The natural calamities (those occurring without the interference of man) studied by the author include *cosmic* events (solar outbursts, falling comets, solar and moon eclipses), *atmospheric* (hailstorms, lightning and thunder, drought, sudden torrential rains, heavy snowfall, tempestuous winds of varying force, such as whirlwinds, tornados, typhoons), and *earthly* events (earthquakes, volcanoes, avalanches).

Nikolay Kolev, Assoc.Prof., PhD, is a Sofia University graduate, Dept. of Bulgarian Philology. He earned his PhD degree from the Dept. of European and Non-European Peoples at Vienna University and since then he has taught ethnology at the University of Veliko Tarnovo. His university textbook "Bulgarian Ethnography" has seven editions. His other main publications are: "*Legends and Myths with Water Sources*" (2000), "*Critical Situations and Traditional Views of the Bulgarians*" (2003, 2005) and "*History of Ethnology*" (2003).

THE PREVENTION AND CONTROL OF BIOLOGICAL DISASTER AND SOCIAL CHANGE (INVESTIGATION STUDY IN THE PASTORAL AREAS IN QINGHAI PROVINCE)

LIANG JINGZHI
PROFESSOR, INSTITUTE OF ETHNOLOGY AND ANTHROPOLOGY, CHINESE ACADEMY OF SOCIAL SCIENCES

Qianghai Province is located on the north-eastern part of Qinghai-Tibetan Plateau, it is source of Yangtze River, Yellow River, Lancangjiang River and other big rivers. It covers a total area of 720 000 km^2 and has a population of 5 000 000. This area belongs to continental plateau climate and is one of the most important pastoral areas in China. Due to unique natural and ecological environment of Qinghai-Tibetan Plateau, natural disasters have been a big problem troubling the pastoral areas in Qinghai. With the transformation of Chinese society after the reform and opening up, especially after grasslands were contracted to each household, the natural disasters in the pastoral areas in Qinghai are obviously in a worse tendency and there are certain new changes about their occurrence rule and type structure, for example, the rampancy of biological disasters like rodent pests is a prominent phenomenon. Then, what is the cause of biological disaster? How is the extent of damage? What are the new changes in disaster? What is the relationship between disaster and social change in the pastoral area? Is biological disaster that has changed the grassland ecology or is the change of grassland ecology that has made the biological disaster worse? Besides natural factors, what influence have the human activities had?

In order to answer the above-mentioned questions, our team chose Henan County and Kuze County in Huangnan Tibetan Autonomous Prefecture with serious natural disasters as target region to conduct field investigation on the natural disasters after the founding of People's Republic of China, especially after 1990s. We use cultural anthropology participant observation and in-depth interview method, and had a preliminary analysis and discussion on the above-mentioned questions according to relevant literature data.

1. The Pests and Biological Disasters in the Pastoral Areas in Qinghai Province

When the number of some biological population reaches high density and breaks through certain scale limitation, this biology is usually marked as pest to be controlled. In the pastoral area in Qinghai Province, the traditional pests mainly include wolves, ochotona curzoniaes, myospakax baikeyis, locusts and caterpillars etc. Because wolves are in the top position of biological chain of grassland and they attack livestock when it is hard or relatively hard for them to find food, wolves have been massively hunted and killed as No1 pest by herdsmen. Especially since the modern weapons like guns were widely used in 1950s and 1960s, the number of wolves population decreased rapidly. Wolves not only seldom make damage or hardly make damage, even the survival of wolves themselves are facing a very difficult situation.

According to our investigation, in Zeku County and Henan County, because China has strengthened the protection of the endangered wolves, prohibit hunting wolves and implementing strict control to guns, accordingly the herdsmen basically have no gun, therefore the number of wolves population is a little bit increasing. The wolves reappear in the pastoral areas that seldom had wolves in the past, and it even has had such an accident that one wolf killed over ten sheep once. Generally speaking, although there are traces of wolf activity, because their total number is limited, they haven't formed substantial harm. In addition, all the herdsmen have the habit of raising dogs, and nearly every herdsman family raises several dogs, which are Tibetan Mastiff entitled as "Deiform Dog" and "King of Dogs". According to the local herdsmen, as long as a herd is guarded by Tibetan Mastiffs, they will not be attacked by wolves. One adult Tibetan Mastiff can guard 400 sheep and defeat three wild wolves and its ability is believed better than a man. However, according to the current actual function, raising Tibetan Mastiff is to prevent more and more thefts rather than to deal with wild wolves. Just as what Duojie, Chief of Yangmu Village, Zequ Town said: "Dogs were used to ward off wolves in the past, but now they are used to prevent thieves. If there is a dog, it will provide a sense of safety to herdsmen."

We can see that the current circumstances of wolves are not optimistic, especially when people discover some medical value from wolves, their fate will be more tragic. Although there is no gun, people still can catch wolves by means of chasing and stalking, etc. It is said that the market price of one wolf is about CNY 1000-2000, one wolf tongue is worth CNY 800 or so, one wolf tail is worth CNY 500 and wolf meat is also expensive, one dogtooth of a wolf can be sold at price of CNY 200. It is believed that wolf meat and wolf tongue have special effects to the cure of stomach diseases, whereas wolf teeth can be used as decoration or amulet. Therefore, when the role of a wolf changes from mischief to victim, its position and function in the biological chain is also reversed. Meanwhile, rodent and insect pests began to stand out.

According to statistics in 2003, among 9 600 000 mu [a traditional unit of area in China] grasslands in Zeku County, 8 000 000 mu grasslands have rodent damage, and the grasslands with serious damage are over 3 500 000 mu. Among 9 130 000 mu grasslands in Henan County, 4 500 000 mu rangelands suffer from rodent damage, and the grasslands with serious damage are over 3 000 000 mu. The rodent pests are mainly ground ochotona curzoniae and underground myospakax baikeyi. It is calculated that one ochotona curzoniae eats over 50g grasses one day, one underground myospakax baikeyi eats over 200g grasses per day. That's to say that feed intake of about 52 ochotona curzoniaes or 13 myospakax baikeyis equals to that of one sheep. Therefore, it is commonly believed that ochotona curzoniaes and myospakax baikeyis are pests that scramble for herbage with livestock and cause grassland degradation, especially myospakax baikeyis, because of their habit of underground activities, in the areas where their population distributes densely, there are many underground caves that crisscross and connect to networks, and on the grounds, there are numerous soil mound heaps that spread all over the place, so myospakax baikeyis are the chief offenders that result in "rat barren wasteland of large area called "black soil beach" by local people. The insect pests mainly include grassland black caterpillars and locusts. The caterpillars in Henan County are most serious, and the disaster area reaches 900 000 mu and it is especially serious in the middle pastoral area of Henan County. The disaster area in Zeku is smaller, but it has also 100 000-250 000 mu each year. After caterpillar disaster, not only the herbages are eaten up, but the places where the grasses are eaten are hard to grow herbages in the next year due to the toxicity left as well. Both Henan County and Zeku County have locust disaster among which the locust disaster in Zeku County is more serious and the herbage volume lost arising from locust disaster is more than 20 000 kg every year.

Apart from rodent and insect pests, the harm of poisonous weeds in the local area become more and more serious, the typical poisonous weeds include fischer euphorbia root, yellow hair crazyweed and Thalictrum alpinum Linn etc., and all of them belong to original grass, most of them are rosette with developed root system and strong vitality. They not only invade the grasslands to contest the survival space with herbage but also often make livestock become addicted and poisoned, even died after mistakenly eating. However, from the seriousness, protracted nature and complexity, rodent and insect disasters, especially rodent disaster undoubtedly is the most important biological disaster in the pastoral areas in Qinghai Province.

2. The Cause of Rodent and Insect Disaster and Social Change in the Pastoral Area in Qinghai Province

In history, although there were rodent and insect disasters in the pastoral areas in Qinghai province, this problem has never been so prominent and serious like today. Then what caused such difficult current situation? We discover during our investigation that many factors influence rodent and insect disasters. Although the natural factors, e.g. insect disaster, climate warming etc. are included, the actions and activities of human beings in the social change are undoubtedly the most important and direct factor. In short, it mainly includes the following aspects:

2.1. The Grassland Contract Breaking Down the Traditional Pasture Rotation System

As the saying goes, each place has its own way of supporting its own inhabitant. Knowledge system is suitable for the grazing tradition in the alpine grassland as well as reasonable use and protection of the grassland in which rotational grazing system is one of important measures that herdsmen take to reasonably use grassland resource and protect grassland ecological environment. After the founding of the People's Republic of China in 1949, even in the people’s commune period, special attention was paid to this tradition, so as to avoid the bad grazing habits like mixed grazing, scrambling for grazing, grazing at random, over-grazing etc. and more scientific and reasonable grazing experiences were summarized and popularized at any time. According to the natural ecology characteristics and herbaceous conditions in different regions, the government divided grasslands into winter and spring grassland, summer and autumn grassland or four-season rotational grazing grassland, and advocated the grazing means of shady slope first and sunny slope later, ditch first and beach later, hill first and beach later as well as grazing at remote places or alpine grasslands, etc. Thus the grasslands not only were fully used but also had necessary rest and growing[1].

After reform and opening up, especially after the grasslands were contracted to each household in the pastoral area of Qinghai Province in 1984, large pastures were segmented into small grasslands with many different sizes, owned by different herdsmen to independently operate and manage. Thus, the traditional rotational grazing system is hard to continue and livestock eat the same piece of grassland in all seasons. Accordingly, the grasslands can't have necessary rest and lose their rejuvenation function. If things go on like this, it will certainly result in the overall degradation of the grasslands. And the degradation of grasslands creates suitable reproductive condition and activity space for rodent pests, especially ochotona curzoniaes. The enlargement of rodent pests' population further accelerates the degradation process of grasslands. In this way, it forms a vicious circle. The research shows that the ochotona curzoniaes have poor eyesight and like dry-touch grassland or short-grass areas with adequate sunlight and sparse grass. The pastoral areas with well-grown grass are relatively wet and forulent, so there are less ochotona curzoniaes and it is hard for them to damage grassland. In addition, traditional rotational grazing method not only provides the rest opportunity for the grassland, but also provides survival condition for the natural enemies of rodents such as wolves, foxes and eagles. The grassland contract system breaks traditional summer and winter rotational razing system and makes the grasslands lose the opportunity of rest, moreover the barbwire fences set up after the contract restrict the activity range of the natural enemies of rodent, and even squeeze their basic survival space. Although currently the herdsmen also carry out so-called rotational grazing in small range on their contract grasslands, but the experiences have proved that such rotational grazing not only can't produce substantial ecological effect but also will interfere with the natural rejuvenation function of the grassland, especially it is prone to induce the spread of rodent pests in big range.

[1] *The General Records of Henan County*, ed. by the Committee of the General Records of Henan County. Lanzhou, 1996, p.335-337.

2.2. Overloading and Overgrazing Accelerate the Degradation of Grassland

To some extent, overloading and overgrazing is one by-product since grasslands contracted to each household. After grassland contract, the herdsmen have autonomy right. So they constantly enlarge the quantity of cattle and sheep in their limited grassland. Although it has once had prosperous scene with cattle and sheep everywhere, because they ignore ecological rules of the grassland itself and limited grazing capacity rule of grassland, as a result, it causes the continuous degradation of the grassland arising from overloading and overgrazing, and further makes the rodent and insect pests rampant.

It is well known that the grazing capacity of a piece of grassland is certain. For example, in Zeku County, approved grazing capacity is one sheep on 8 mu land, but in fact, it reaches 12-13 sheep. In Henan County, according to the introduction of Mr. Peng Mao, Master of Grassland Management Station, in 1970s, although there are areas with ground rodent disaster in the whole county, the disaster area is small. Until 1980s, the grassland situation was still very good and the landscape was very beautiful as well. In the last decade, especially after grassland contract, rodent disaster began to really emerge and became increasingly serious because of overloading and overgrazing. Overgrazing makes the grazing capacity of grassland to be beyond the limitation and aggravates the loading and ecological pressure of the grassland, accordingly it results in "livestock and rats scramble for grass" and "the contradiction of human beings and rats". In fact, in order to keep the survival of their population, rodents themselves also know how to protect the environment of their survival land and conduct timely and reasonable adjustment to their population number. But the grassland degradation arising from overloading and overgrazing has deteriorated the survival environment of rodents. In order to adapt to changed natural environment, the rodents have to scramble for grass with human being and livestock at the price of objectively damaging grassland. Meanwhile, because of overgrazing, after herbage is eaten up, poisonous weeds have room of expansion, and the spread of poisonous weeds provides more activity room for ground rodents. Moreover, the underground rodents subsequently become more active because there is a relationship of mutualism between the ground rodents and underground rodents. In addition, the increase of herds will produce large quantity of dung, especially cattle dung. If excessive cattle dung can't be timely picked up, they can also burn herbage to death. So, besides rodent pests, the formation of dark soil beach has certain relation with excess cattle dung and the centralized grazing and trampling have certain relation, but the most fundamental reason is the grassland degradation arising from overloading and overgrazing.

It is notable that in the settlement process brought by grassland contract, it gathers objectively more population in certain space, the population gathering together with the permeation of modern life style to the pastoral areas have following influences: a) producing plenty of domestic waste, while in the past, there was no such waste; by using the words of Wan Decai, vice director of Henan Agriculture and Grazing Bureau, it is that the herdsmen have no culture of producing waste in their traditional living habit - for example, if they want to go to toilet, they will do it in a rat cave and there will be no waste with the function of natural weathering, and b) the formation of non-natural population gathering, which usually makes the grasslands in and around the settlement being trampled for years, so degradation of grassland is accelerated and the settlements become actual dark soil places.

2.3. Uncontrolled Hunting and Poisoning Acts Result in the Massive Death of the Natural Enemies of Rodents and the Breaking of Biological Chain and Loss of Ecological Balance

Hunting has been a traditional auxiliary economic activity in the pastoral areas and the wild animals with high economic values like wolves, corsacs, lynx and woodchucks usually become the main objects of hunting. According to the Records of Zeku County, only in 1980, the whole county purchased 30 943 woodchuck skins, so we can see how active the hunting is. Since the late 1980s, the government advocated to protect wild animals and prohibited hunting by explicit order, but after large-scale hunting for many years, the population number of wild animals like wolves and corsacs etc. has decreased dramatically; some wild animals, e.g., lynx and snow leopards etc. are on the edge of extinction. Herdsmen took particularly poisoning rodents actions in large area in the pastoral areas for long time, from point of views of the actual effect, it hasn't achieved the goal of killing rodents, instead it makes the natural enemies of rodents, such as wolves, corsacs, weasels and haws and falcons and other zoophages massively dying because of mistakenly eating the poisoned rodent corpses. Ultimately it leads to the breaking of biological chain, and the rodents, who have natural resistance against chemical medicines, lose natural enemies, and obtain very big survival space. So their population rapidly increases and forms inextinguishable rodent disaster. For example, in Henan County, even in 1970s, the natural enemies of rodents such as corsacs, weasels and haws and falcons were found everywhere in the grassland around the county town and the rodents in the pastoral areas were not rampant as nowadays. After then, from 1980 to 1992, because hexachlorocyolohexane, a kind of chemical poison bait was widely and constantly used to kill rodents on a big scale. As a result, although it won the victory to the rodents for a time, meanwhile it also caused massive death of the natural enemies of rodents like wolves, foxes, weasels, haws and falcons, even some livestock were poisoned to death. In 1992, Henan County was honored as “County without Ground Rodent Pest” and it is also the first county that has no ground rodent pest in the whole country. Happy days did not last long, the rodent pest reappeared in 1993 and became a disaster in 1995, the rodent pest in 2003 was the most serious and even it can't be effectively controlled nowadays.

2.4. Population Pressure is the Objective Factor of Grassland Degradation

Mr. Suonan, the County Magistrate of Zeku County thinks that the grassland area in Zeku County and in Henan County are nearly the same, but the total populations have big difference. At present, the population in Zeku County reaches 60 000, but in Henan County population is only 30 000, so the contradiction between people and grass in Zeku County is more serious, accordingly the grassland degradation and rodent pest are more serious. According to statistical data, the total population in Zeku County in 1953 was 17 254 and the population in 2nd National Population Census in 1964 was 18 504, and population in the 3rd National Population Census in 1982 was 34 330, which was 15 826 more than that in 2nd National Population Census and the increase rate of population was 85.53%, and the population density was 5.16 persons/km^2. In the 5th National Population Census in 1990, the total population in the whole county was 42 489, which was 8159 more than that in the 3rd population census and the increase rate was 23.77%, and population density was 6.38 persons/km^2. That is to say that the current total population in Henan County is 4000 persons less than that in Zeku County in 1982 and the population density is less than 4.5 persons/km^2. It shows that Zeku County is facing severer population pressure and the contradiction between people and grass is sharper. To some extent, it can explain why the rodent pest in Zeku County is more serious than that in Henan County.

2.5. Uncontrolled Gathering and Digging, Destroying Grassland Vegetation Causing Grassland Degradation and Desertification

Traditionally, gathering industry is a necessary and complementary means of the animal husbandry economy in the pastoral area, wherein the gathering of medicinal plants, such as Chinese caterpillar fungus, Chinese rhubarb, large-leaved gentian and snow lotus, play a major role. Let's take Zeku County as an example, the whole county purchased Chinese rhubarb of 161.95 tons in 1960, large-leaved gentians of 137 tons in 1969, notopterygium incisium of 22.74 tons, and Chinese caterpillar fungus of 123.5 kg[2]. Due to rich wild plant resource at that time together with cheap price, and the medicinal plants were uniformly purchased by the national medicinal material department, therefore, generally speaking, the gathering and digging activities didn't have obvious influence to ecological environment. But in recent years, due to the soaring price of rare medicinal herbs like Chinese caterpillar fungus, the excessive gathering and digging phenomena is more and more serious, the environment problems and civil disputes caused thereby becoming more and more serious. Although the local governments prohibit auction, sublease and assignment of grasslands with Chinese caterpillar fungus by explicit order, buying, selling or assignment under disguise are still common. In order to pursue profit maximization, renter and leaseholders merely gather and dig without control, local people seldom consider the grassland protection and sustainable utilization, accordingly it directly resulted in the destruction of grassland, and further caused degradation and desertification of partial grasslands; it also brings out the serious rodent and insect pests as well as the spread of poisonous weeds.

2.6. Unplanned Reclamation and Cultivation Aggravating the Process of Grassland Desertification

Promotion by national vigorous development of agriculture, Henan County blindly pursued the policy of "reclamation is the key link" and organized Youth Reclamation Team to cultivate the grassland and establish

---

[2] *The General Records of Zekou County*, ed. by the Committee of the General Records of Zekou County. Beijing, 2005, p.174.

farms. A total area of over 200 000 mu grasslands was subsequently changed to farms, which accounted for 1.87% of the usable grasslands of 10 670 000 mu. In 1960, 71 940 mu lands were seeded, but the seedlings in 21 682 mu lands didn't come out, and 29 878 mu lands were hit by a natural calamity. Because it is hard to grow crops in an alpine region and developing agriculture in those areas are against natural law, so reclamation and cultivation were stopped at the end of 1961, but the damaged grasslands still can't restore to the original state even after 20 years.[3] The situation in Zeku County was more serious than that of Henan County. Since 1958, Zeku County began to reclaim lands and grow crops and establish farms at a trial basis in Heri Township, Ningxiu Township, Wangjia Township etc. where there are relatively good natural conditions of climate, soil, water resources etc. From 1972 to 1982, Zeku County began to migrate to develop agriculture and widely plant crops like highland barley, oats, rapes, wheat and potatoes etc. The planting area of rapes was increased from 0.2 hectares in 1958 to 3670 hectares in 1995. Unplanned reclamation of grasslands not only results in degradation and desertification of grassland, but brings out serious rodent and insect disaster as well. The facts show that the areas where reclamation was gathered are also the places with the most serious rodent and insect pests. Just like what Master of Huangnan Grassland Work Station said: "The serious desertification in the western Zeku causes the degradation of the grasslands in big area, which mainly resulted from blind reclamation. Furthermore, due to over grazing, the food fighting of livestock and rodent pests bring out big-area black soil beaches. Now the county has over 600 mu lands with high degradation and over 3 000 000 mu lands with middle degradation. Reclamation of waste lands and rodent pests are the two major factors. Although now the state invests to conduct the restoration of the reclamation lands and make effort to restore to original ecological environment, the restoration difficulty is very big."

In conclusion, the cause of rodent disaster in the pastoral areas in Qinghai has direct relationship with the social change factors. The grassland contract system implemented since 1980s terminated traditional rotational grazing system and made the grasslands lose the rest opportunity and rejuvenation function. Furthermore, grassland degradation resulting from the factors like overloading and overgrazing, population pressure, uncontrolled gathering and digging, as well as the massive death of natural

[3] *The General Records of Henan County*, ed. by the Committee of the General Records of Henan County. Lanzhou, 1996, p.87-88.

enemies of rodents from second poisoning make the population number of rodents constantly expand. As a result, rodents scramble for grass with human beings and livestock, which further aggravate the degradation of grasslands and form a vicious circle of grassland degradation-rodents overrunning-re-degradation, but the most fundamental and important factor is the activity of human beings. Therefore, we would say that grassland degradation brings out rodent pests rather than rodents cause the degradation of grassland, rodent pests are the result of grassland degradation instead of a reason for it.

3. The Prevention and Control of Rodent and Insect Pests and Social Change

The prevention and control of rodent and insect pests depend on the development level of science and technology, meanwhile it also depends on the nature of rodent and insect disasters and the recognition to their laws. It can say that people's knowledge on rodent and insect pests has experienced a tortuous and complicated process. As a consequence, the prevention means should also be different in different times, and have different characteristics in different times. Generally speaking, the prevention and control of rodent and insect pests have experienced three stages, i.e. 1950s-1970s is the first stage, 1980s-1992 is the second stage and from 1993 to present is the third stage.

In the first stage, it completely used chemical medicine prevention and control method, and used medicines like Zinc phosphide, zinc fluoride, Fluoroacetamide to kill rodents. In 1965, a rodent killing experiment was conducted by plane in Zeku County in spring and nearly 50 km^2 lands were spread with poison baits. After sampling, the rodent killing rate averaged at 93.5% and had good result once, but due to insufficient knowledge about the reproductive law of the rodents and their drug resistance capability, usually the rodent pests can restore to the level before the killing within three to five years. However, due to the fact that ecological environment in the pastoral areas then was generally in good condition, and the population was small, while the grassland area is relatively broad, and rotational grazing can be taken, therefore, the rodent and insect pests didn't develop to an uncontrollable serious level, the disaster area and disaster intensity was much less.

The second stage is to continue to use chemical medicines and further reinforce the killing to rodent and insect pests on a big scale and in large range and full aspects. For example, in Zeku County, from 1980 to 1990, by continuous killing, the accumulative areas of rodents prevention reached over 1000 km^2 and the average killing rate is 93% or so. Meanwhile, killing locusts by mechanically chemical spray was implemented and got good effect for once. Using new researched 2.4—DJ vinegar medicine solution to deal with poisonous weeds like fischer euphorbia roots etc, took effect in the same year, but poisonous weeds grew better in the next year. After 3-year experiences, the effect was not good. From the surface, killing effect especially to rodent pests was very prominent and a great number of rodents were killed, and both Henan County and Zeku County were granted "County without Ground Rodents". But meanwhile second poisoning phenomena was very serious, the natural enemies of rodents like wolves, corsacs, weasels and haws and falcons successively died because of eating the corpses of rodents, and a great number of pesticide residues not only polluted the environment of grasslands but also even caused the death of livestock like cattle and sheep. More seriously, this effect didn't last long, and it was encountered with a desperate counterattack from rodents. The killing actions on a big scale only resulted in the temporary reduction of rodents. After 1993, the rodent pests became more and more serious, nearly reached an uncontrollable level.

The third stage, with the awakening of environmental protection consciousness and reinforcement of ecological concepts, the people began rationally self-examining the environmental price and its unfavorable consequences of purely using chemical medicines to kill and control rodent and insect pests, and attempted to find a safe and reliable means to achieve the object of killing and controlling rodent and insect pests. Therefore, the people began to give up the control method by using chemical medicines, and use newly-invented biotoxin to deal with rodents (ground rodents). But what the people worried is that although this medicine currently has no direct and obvious harm to herdsmen and livestock, after all it is a kind of toxin and its potential influence is indispensable, after using for long time, it may bring some unfavorable effect. In addition, in some areas, the people attempt to use biological control technology, such as setting up scaffolds to attract eagles, introducing weasels, to prevent and control rodent pests. But this method has some limitation, what eagles catch are nearly the old, weak, sick and disable rodents, and it can't eliminate rodent pests fundamentally. Nevertheless, this method has positive function to consolidate the killing effect. When killing work achieves some effect, using this method as its supplementary can play the role in controlling rodent population and consolidating the effect. It is introduced that one adult eagle can catch 3 to 5 rodents one day and one weasel can catch 5 to 7 rodents one day. To control locust pests by raising specially-trained chickens is also regarded as a good method and its economic effect is also good. It is said that drivers that pass by like to buy this kind of chickens because they are deemed as a green product.

It should say that biological control technology is a method with good prospect. But from the current situation, there is very big limitation and it is not realistic to popularize in wide range, moreover, to restore the biological chain that has broken is not a thing that can finish in short time. Meanwhile, we have to acknowledge that currently the research on the habits of ochotona curzoniaes, especially their position and function in the biological system of plateau hasn't been paid enough attention, but it will determine the people's action and attitude to ochotona curzoniaes to great extent. That is to say that the crucial reality requires us that scientific research must sincerely and truly answer the biological function and position of ochotona curzoniaes and truly understand and know such specie.

Regarding the prevention and control of underground rodent pests, the main method that is used is physical method, which is to use bow and arrow rodent trap to conduct artificial catching. Because this method is simple and practical, only in 2006, Henan County caught 220 000 underground rodents and Zeku County caught 230 000. Because the rodent skins can replace the medicine value of tiger skins in curing rheumatism, dead rodents after peeling off the skin and getting rid of meat are uniformly purchased by Tanggula Pharmaceuticals. Bow and arrow rodent traps are provided by the local grassland station, and would be taken back after using. The implementation of rodent catching is uniformly organized by the grassland stations and carried out by special rodent catching teams. It is paid as per mu. Meanwhile, the herdsmen are trained and instructed on rodent catching techniques. The governments provide certain rewards for those that have reached deratization standards, for example, Henan County awards one Yuan for one mu. If all the lands have reached the standard, each mu will be subsidized 0.4—0.5 Yuan for a second time.

In addition, before massively using improved iron bow and arrow rodent traps, there were some traditional prevention and control method to underground rodent pests in the folk, such as blowing air into the underground rodent cave through a sheepskin tube at one end and setting iron clamp at the other end; or stopping up a rodent hole with mud, stones and cattle dung after lunch in winter, at night, the mud would be frozen together with rodent hole. In this way, the rodent hole was stopped up and the rodents can't come out of the hole and would die finally, or pouring rodent holes with water, or using throwing-stone machine, or beating with a slingshot. Although these methods had certain effect when the underground rodents were not many in the past, when underground rodents overran, these methods become entirely inadequate. Therefore, the most important method used nowadays is bow and arrow.

Traditionally, the simplest method for herdsmen to deal with insect pests is to burn partial grasslands to kill worm eggs in winter. When the insect pest doesn't exist in large area, the disaster mechanism is relatively simple and the grassland is relatively broad, this method is a good one. As the insect disaster is becoming more and more serious, frequent and the reason for it is complicated, it seems that there is no more effective killing method other than chemical medicines. According to Master Peng Mao of Henan Grassland Station, most insect disasters happen in the dry season in spring, due to the suddenness and locality of disaster causing means, therefore, once it breaks out, it is hard to control. Currently the main method is to use pharmaceutical spraying to kill and control insect pests and the best time for killing and control is only one week or so. Once it is missed, it would be hard to have efficiency. So it needs a great deal of manpower and materials and a lot of money, to employ some manpower and the grassland station to provide technical support and medicines. But in principle, the herdsmen are encouraged to kill and control pests by themselves. Generally speaking, one man employed for one day can kill and control insect pests in 20 mu grasslands, while if the killing and controlling are done by herdsmen, it can reach about 7-8 mu. Even so, because killing and controlling by herdsmen themselves is more careful, the effect is much better. An employed worker is paid as per mu, killing and controlling insect in one mu is 0.4-0.5 Yuan. Because the state provides subsidies as per the area, the grassland station also calculates payment as per the area. As to poisonous weeds, there is no better prevention and control method; basically they are removed by manpower.

Comparing with Henan County, the pests in Zeku County seem more severe. According to the introduction by Wandecairang, Chief of Henan Agricultural and Animal Husbandry Bureau, generally speaking, the insect pests in the county are becoming more and more serious, and there is no better prevention and control method except chemical medicine. In the People's Commune Period, the pesticide funds were all provided by the country, but now it adopts self-raised funds together with national subsidiary to buy medicines to kill insects. The pesticide works are mainly done by the employed people from Han nationality from inner land with spray-ers. But anyway, the residues of medicine have some harm to livestock like cattle and sheep etc. The pesticide is usually done in spring because the insect pests reproduce in spring, become disaster in summer, and hibernate in winter. One insect can eat a lot of grass in one day. When they become disaster, they are close and numerous just like sheep dung in sheepfold. In the places where herbages have been eaten up, they left toxicity and there will be no herbages in the next year. When some herdsmen pick up insects with plastic bags, their hands become inflamed because of poisoning. Although the disaster of palmerworms is generally partial, they come fiercely. Generally speaking, the places with insect pests are the places for pesticide, especially in droughty years, there are such a lot of palmerworms, but in years with more rain, usually there are much less palmerworms.

In addition, it should be acknowledged that although the relevant government departments like grassland stations, animal husbandry bureaus etc. play an important role in the organizing and coordinating herdsmen to prevent and control the disaster works such as rodents and insects pests, due to the change of land system, operation mode and herdsmen's concepts after grasslands contract, each family does things in their own way and people began to produce and act around their own family, and the daily communication between herdsmen are obvious weaker than before, therefore, the pest prevention and control can't be organized, coordinated and implemented as easily as before, the unified activity of collective pesticide is also gradually replaced by the pesticide conducted be each family. In this way, because of losing the unified activity of collective pesticide, it can't thoroughly stop the trans-regional move and spread of rodent and insect pests, accordingly it is hard to achieve and consolidate the prevention and control effect, while the prevention and control work objectively require unified action in unified time, we particularly should do so when dealing with insect and locust pests. Of course, in the inhabited pastoral villages (e.g. village inhabitation places), it is easy to organize the action because of population concentration, so the prevention and control work is organized by the village to unified implement or coordinated by the village to be conducted in scattered actions by each family or each team.

It should say that it is the activities of human beings that have changed the original ecological relation and broken the biological balance; meanwhile, it has also changed activity modes of all other animals. Moreover, the other animal populations who have had their activity modes changed also influence the activities of human beings and their traditional values. This is a dialectical and interactive process. Although as a nationality that all the people believe in Tibetan Buddhism, ahimsa is one religious discipline that they should strictly abide by. But now since the human beings and rodents are combating for grass and the contradiction between human beings and rodents are sharper, poisoning rodents is their helpless choice to defend their own interest and right of existence. So called "Helplessness" is to say this is not an effective and fundamental solution. From the standpoint of Buddhists, the viewpoints of Jiumeixiazhu, a monk from Heri Temple, Heri Township, Zeku County and Director of the Temple Management Committee maybe have some representativeness: "From the perspective of a monk, killing rodents is not a good method. Moreover, it can aggravate our own sin. But if we don't kill rodents and prevent rodent disasters, it doesn't accord with the policy of the government. In fact, even during the World War I and World War II, human beings killed many rodents with poisons, but the result is that there are more and more rodents. We can say that rodents can't be killed out. In the view of Buddhism, the formation of rodent disasters and dark soil beaches are all the sins committed in the previous life and nemesis. Regarding locust disasters, the living Buddha and NGAG-PA in the temple are often invited to recite scripture and practice magic, so as to drive locusts away. " Obviously, although the standpoints are different, the people from all the circles of the society have reached a basic consensus that simply using poisoning method can't fundamentally solve the contradiction between human beings and rodents.

In conclusion, under the background of social change, the frequent rodent and insect pests in the pastoral area in Qinghai is only the product of ecological environment in nature, and it is an expression form of ecological disaster caused from improper activity of human beings. Therefore, the prevention and control of rodent and insect pests in the pastoral areas still need to start from the height of comprehensive control of ecological environment, feasibly adopts measures such as reducing livestock and limiting grazing, determining the number of livestock according to the grass condition or converting the grazing for pasture, returning farm for pasture, enclosing and culturing grassland and ecological operating, so as to realize the harmonious development of human, society and nature. Only in this way, can it fundamentally solve the historical difficult of rodents, insect and other biological disasters.

4. Concluding Remarks

In this paper, we conduct a preliminary investigation on the construction and characteristics of biological disaster as well as the cause of biological disaster, prevention and control methods and the correlation between it and social change. By investigation, the following conclusion can be made: the biological disasters in the pastoral areas in Qinghai traditionally manifest wolf disaster, rodent disaster and locust disaster, but now it is the rodent disaster and insect disaster, and the rodent pests which are the most serious. The occurrence law and the change of characteristics of biological disasters have obvious internal correlation with current social change. Since reform and opening up, the grassland contract system implemented in the pastoral areas broke down the operating mode in the past, big pastures are divided into small grasslands, traditional rotational grazing system can't continue. The increasing population, the overloading and overgrazing and unreasonably medicine used are the fundamental reasons that cause the grassland degradation, desertification, biological chain breaking, and increasing deterioration of ecological environment. In other word, the biological disaster is the man-made disaster in nature and the improper activities of human beings are the main reason that causes the overrunning of rodent and insect pests.

Therefore, from current situation, the key to the solution should first take the following methods into consideration: a) to deepen and improve grassland contract responsibility system, discuss and explore the approach or pattern that can better combine grassland ownership and use right, attempt restoration possibility of the traditional rotational grazing in big area, and adjust the unbalance between herdsmen with less livestock and those with more livestock in grassland utilization by compensable transfer or other compensation mechanism; b) to establish and strengthen ecology compensation system, practically solve the worries of herdsman after converting the grazing for pasture and stimulate the herdsmen's initiative of doing so, limiting grazing, even prohibiting grazing; c) to strengthen the scientific research to plateau ecosystem, especially the biological population, from the height of keeping and maintaining biological diversity, properly understand all kinds of biology, especially the position and function of so-called pest population like ochotona curzoniae, myospakax baikeyi in plateau ecosystem[4]; d) to continue to increase the scientific re-

[4] With the development of modern biology, the academic circle constantly deepens the knowledge of biodiversity. One point of view thinks ochotona curzoniaes,

search force, make unremitting effort to develop and popularize more diversified biological control technology, such as setting up scaffolds, introducing weasels and corsacs. Although their effects are not big and it is hard to popularize them in big range, their development prospect should not be doubted. Meanwhile, accelerating the research and development of safe, reliable and controllable biological preparation without toxicity and side effect is also a makeshift to restrain excessive reproduction of rodents at present. In addition, to increase the protection to the natural enemies of rodents, like wolves, prohibit hunting in any form, widely develop and strengthen propaganda and education and create a good social public opinion atmosphere are also very important. Practice has proved that only if ecological environment protection concepts are deeply rooted in the heart of the people, can they become conscious motivation of the peoples' action.

References

Lai, Chien-Hsun and A.T. Smith. The Key Position of ochotona curzoniaes: the Influence to the Birds Biodiversity. – In: Wang S. et al (ed.) *Protect Biodiversity of China*, Vol.2. Beijing: China Environment Science Press, 1996.

The General Records of Henan County, ed. by the Committee of the General Records of Henan County. Lanzhou, Gansu people's Press, 1996.

The General Records of Zekou County, ed. by the Committee of the General Records of Zekou County. Beijing: Chinese Annual Review for Records of County and Town Press, 2005.

---

which is commonly regarded as pests are actually one important part of plateau ecosystem. They not only play the role of "heterogenic engineer", but also is key specie of plateau biodiversity. Therefore, Poisoning ochotona curzoniaes on large scale is one factor that results in the losing of grassland function. (See: Chien-Hsun Lai and A. T. Smith. The Key Position of ochotona curzoniaes: the Influence to the Birds Biodiversity – In: Wang S. et al (ed.).*Protect Biodiversity of China,* Vol.2, 1996.

Illustrations

1. Desertization of grassland in Zeku County

2. Herder, sheep and degrading grassland in Henan County

Abstract

The bio-disaster refers to the disaster caused by locusts, rats, wolves etc. The bio-disaster of Qinghai Pasture mainly shows in the form of insect disaster, mouse disaster and the wolf disaster. Owing to the natural and cultural particularity of pastoral areas in Qinghai, the bio-disaster has regional, contemporary and ethnic features. The wolf and locust disasters were the most severe in the past, but now the disaster of mouse and insect takes the place, and the mouse disaster is more serious. There exists inherent relevance between the law and characteristics of the bio-disaster and the changes of current society. Since the reform and opening–up in China, the old collective production system has been abolished because of the family contracting out system in pastoral areas. The wide pasturelands have been divided into small ones with various striped areas. Consequently, the traditional rotation system in summer and winter cannot be continued. With the increase of population, the over grazing and unreasonable dividing of grassland are the root causing to the degeneration and desertification of the grassland, the breaking of bio-chain, the flood of mouse and insect disaster, and the worsening of grassland ecological deterioration. In other words, the bio-disaster is an artificial disaster essentially; human beings' non-proper activity is the chief culprit in the rampancy of mouse and insect disasters.

Liang Jungzhi, Ph.D., Professor at the Dept. of History, the Institute of Ethnology and Anthropology, CASS. His scholarly interests are focused on the history of Chinese root religion and the history of disaster in China's ethnic communities. Representative publication: *The Root religion in Qing Dynasty and Local Society (2004).*

FOOTBALL GAME DURING AND AFTER A CRISIS PERIOD: CASE STUDIES FROM BULGARIA

IVA KYURKCHIEVA
ASSIST.PROFESSOR, ETHNOGRAPHIC INSTITUTE AND MUSEUM, BULGARIAN ACADEMY OF SCIENCES

The social nature of sports, football in particular, is immanently related to the development of contemporary society. Sport is part of the course along which political, social and moral models evolve. It reflects some of the processes in the social environment and culture, and it can serve as an important source for their analysis.

The current text presents a view on football and related activities as part of the social developments in Bulgaria during the last decade of the 20th and the first decade of the 21st century. The social, economical and political crisis after the changes in 1989 gradually affected all aspects of everyday life. Only recently – in the past 4-5 years, there is an emerging public sentiment that the crisis is being slowly overcome. The research presents two separate examples of football related activities and emotional outbursts among the fan base of two village football clubs. They are studied as part of the events resulting from the overall crisis.

The first example deals with the football club of the village of Galata (Teteven region in North Central Bulgaria), whose greatest achievements were in 1997 (Illustration 1). My observations of the social processes, related to the success of this local team, are from the same period. This period marked the peak of the impact of the national scale economic and political crisis. In this case, football can be studied as a type of reaction to the crisis and a major factor, accompanying the attempts to deal with it.

The second example presents the revival of several village clubs in the Garmen Municipality (Gotse Delchev region in Southwest Bulgaria) in 2004-2007. In this case football is studied in the context of coming out of the crisis, since this was a period of gradual economic and social stabilization, which could be sensed in the everyday life. The football enthusiasm in the mentioned municipality could be interpreted as a milestone, marking the turning point in overcoming of the crisis.

Football has been important for the preservation and expression of the identity among certain groups of people. In this connection it has been very important for the preservation and demonstration of the cultural identity of communities, which construe themselves as different from the majority in the state. The common premise for the two case studies is that they are set in Bulgarian Muslim (so-called Pomak) communities – the population in the village of Galata is fully comprised of Bulgarian Muslims, and the predominant majority of the people in the Garmen municipality also belong to such community. Living there are representatives of different ethnic and religious groups (56.31% Bulgarian Muslims, 30.05% Bulgarian Christians, 6.94% Roma and 6.7% Turks)[1]. In the area of Teteven, Bulgarian Muslims live independently in three villages (Galata, Gradeshnitsa and Glogovo), and in a mixed community – in Bulgarski Izvor, Toros and Kirchevo, whereas the surrounding population is mostly Bulgarian Christians. For the Bulgarian Muslims in both regions football has proven to be a means of emphasizing their local identity[2]. The reaction of the local population to the others, what separates them or integrates them with the others can be interpreted through the prism of football.

Besides the common trends within the context of the ethno-religious interactions with the neighbouring population, football has also proved to be an opportunity for display of different political preferences. The emotions along the line of politics and football prove to be very important in both of the studied locations. The amalgamation of the local authorities with the management of the football clubs and the supporters is part of the everyday life and is a source for excitement in the community. The rise or decline of local politicians is often dependent on their attitude towards the activities and achievements of the local teams.

The interpretation of sporting events, including football, through the viewpoint of ethnological and anthropological analysis has long been present in specialized literature both abroad[3] and in Bulgaria[4]. The basic

---

[1] The information about the municipality comes mostly from its website (http://www.garmen-bg.com) and Strategy for the Development of the Municipality of Garmen for the Period 2007-2013.

[2] For greater details see: I. Kyurkchieva. Svetat na bulgarite myusyulmani ot Tetevensko – prehod kam modernost. [The World of Bulgarian Muslims from the Teteven Region – Transition to Modernity]. Sofia, 2004.

[3] A.Cheska. Sport Spectacular. A Ritual Model of Power. *International Review of Sport Sociology*, Vol.14, 1979, p.51-72; R.Lipsky. Towards a Political Theory of American Sports Symbolism. – In: J.Harris, R.Park, (eds.). Play, Games and Sports in Cultural Contexts. Champaign III, 1983, p.79-91; J.Hargreaves. Sport, Power and Culture. A Social and Historical Analysis of Popular Sports in Britain. New

methodology in this work has been the comparison between oral field research information, predominantly semi-structured interviews with representatives of different gender and age groups, and written materials from the specialized sports media and local press. This approach has proved its effectiveness in helping to obtain a more complete overview of contemporary life. The study dedicated to the Bulgarian Muslims of the region of Teteven is based on ethnological materials collected during several expeditions in the period 1997-2000[5]. The interviews in the Garmen Municipality were conducted in 2007-2008[6]. They reflect well the current socio-economic development and the football enthusiasm in this region.

Dimensions of the Crisis

In the beginning of the transition in the 1990s, unemployment was the main problem for the people in Bulgaria because until 1989 they were mostly employed within some branch of the centralized socialist economy. For the Bulgarian Muslims of the Teteven region, after a period of total

York, 1986; E.Wagner. An Overview of Sport in Asia and Africa. – In: E.Wagner (ed.) Sport in Asia and Africa. Comparative Handbook. New York, 1989, p.3-11; E.Hobsbawm. Mass-Producing Traditions: Europe 1870-1914. – In: E.Hobsbawm, R. Ranger (eds.). The Invention of Tradition. Cambridge, New York, 1992, p. 263-307; I.Cholovich. Football, Hooligans and War. - In: N. Popov (ed.), The Road to War in Serbia, Trauma and Catharsis. Budapest: CEU, 2000; D.Booth. Escaping the Past? The Cultural Turn and Language in Sport History. *Rethinking History*, Vol.8, No1, March 2004, p.103-125.

[4] P.Petrov. Praznichni tradicii i politicheski praktiki. Dissertation Thesis, BAS, Sofia. [Festive Traditions and Political Practices, PhD Thesis, Institute of Folklore at the Bulgarian Academy of Sciences], 2002; B.Georguieva. Sportat v bulgarskia grad (kraya na XIX – sredata na XX v.). [Sports in the Bulgarian Town, End of the 19th–Middle of the 20th Century]. – In: *Godishnik na Sofijskia Universitet* [Yearbook of the Sofia University]. Vol.93-95, 2000-2002. Sofia, 2003, p.69-118; M.Iliycheva. "Faithful Until Death": Sports Fans and Nationalist Discourse in Bulgarian Internet Forums. Budapest, 2004; I.Filipova-Kyurkchieva. Football and the Political Symbolism in Bulgaria in the 1980s and 1990s. *EB*, 10, 2006, p.345-362.

[5] The 1997-1999, expeditions were carried out thanks to sponsorship of the International Center for Minority Studies and Intercultural Relations (IMIR) Sofia. The 2000 expedition was supported by the Research Support Scheme of the Open Society Support Foundation, grant No.: 993/2000.

[6] The information has been collected during my participation in the IMIR project "Minority Rights and Inter-Ethnic/Inter-Cultural Relations in Mixed Population Municipalities", sponsored by the Central and East European Trust (CEE Trust). I gratefully acknowledge the cooperation of all my colleagues on this topic.

unemployment, one of the possible variants of finding employment was working abroad. In the early 1990s a great number of young men (construction workers mostly) found employment in Germany. However, this was seasonal work (only during the summer period) and after 1996, projects demanding such specialists were becoming increasingly rare. In the summer of 1997 the economic crisis intensified and a small share of the local population was employed in construction of private projects in Serbia.

In the 1990s continued the socialist period tradition, whereby local young men found an opportunity for self-realization in sports. Another part of the population continued to support themselves by activities, which had been traditional in the recent past - the timber industry continued to be a substantial source of income and almost all the families in the villages kept livestock and had opportunity to sell the surplus milk.

During the second half of the 1990s, women found it increasingly difficult to find employment. It was usually minimum wage and involved unskilled labour. There were a few cases of women going abroad to work as nursing assistants or housekeepers. However, the Bulgarian Muslims from the region of Teteven remained personally committed and had an active approach towards the developments taking place in the country as a whole.

The situation of the population in the Garmen Municipality was not much different in the 1990s. Some people relied on humanitarian assistance in the hardest years of the transition. The crisis, which adversely affected the Rhodope Region, subsided in the period after 2004. Currently, the predominant impression of the municipality is one of vitality and activity of the people[7]. Men are engaged in the construction industry in Bulgaria and abroad, whereas the textile and footwear production facilities are operating again after a long stagnation period and provide employment to some of the women. People continue to be engaged in agriculture. The cultivation and processing of tobacco, although dwindling, supports the livelihood of many families. The opening of the Ilinden Border Checkpoint and Bulgaria's accession to the European Union has provided access to seasonal farm work and other temporary employment in neighbouring Greece. At the same time the mineral springs and the existing resort compounds and family hotels around the village of Ognyanovo, the proximity of well known resort settlements and historical and ethnographic reservations like Bansko, Kovachevitsa, and Leshten, make the municipality the object of intensively developing rural and cultural tourism. This picture of

[7] Unemployment decreased from 38% in 1998 to 8% in 2008.

dynamic processes in the present day has been an important condition for activity in the sphere of sports.

Football during the Crisis

For the Bulgarian Muslims of the region of Teteven sports have provided accessible way of spiritual expression, community projection and presentation to "the others"[8]. Common interest in sports is, to a certain degree, the result of actively operating kinship relations in the present, and, on the other hand, it intensifies the unity and solidarity of the group. From this point of view, it may be claimed that among the Bulgarian Muslims of the Teteven region interest in football has developed parallel to the crisis and is part of its manifestations. The activities related to the sport have also provided opportunities for employment.

By the middle of the 1990s, the Olympic-Galata Football Club was established owing to the zealous efforts of several generations of community members with an athletic background[9]. Originally, the players were recruited exclusively among the local youth. Last, but not least, it should also be mentioned that the local football teams were able to continue their existence due to the financial backing of local wrestlers, who were famous athletes in the 1980ies and subsequently set up their own businesses, including private security companies. It could be claimed that this was the point from where on certain people amassed economic wealth, often illicitly. These processes are typical for a period of crisis, but due the initiative and support of such individuals, the local children received scholarships for professional sports instruction (usually in wrestling or martial arts) in Lovech and Pleven (Central Bulgaria).

---

[8] The example presented about the Bulgarian Muslims of the region of Teteven is part of an article, dedicated to kinship and solidarity in the group – I.Filipova-Kyurkchieva. Kinship and Solidarity Among the Bulgarian Muslims of the Teteven Region. *EB*, 5, 2001, p.139-150.

[9] The local population has had famous representatives in wrestling under the Ottoman Empire. Under socialism, a great number of famous wrestlers, who won national and international competitions, were from the Bulgarian Muslims of the Teteven region. The traditional wrestling at that time (according to P.Petrov) continues to be an outlet for local or regional culture, but has been reproduced in conformity with the regulations, prescribed by the authorities. P.Petrov. Koi, kak, za kakvo se bori. Za niakoi upotrebi na "narodnata borba', [Who Wrestles, How and for What? On the Utilization of 'Folk Wrestling' in Bulgaria], *BE*, 2000, No3, p.32-35.

The local authorities in the village of Galata, personified by the mayor, were also involved in the activities of the football club. The club sponsors had even pledged to restore the healthcare services and infrastructure of the village.

Another positive circumstance for the development of the club was that the village of Galata already had a football stadium, which was built during the socialist period.

Utilising all these favourable conditions, in 1997 the team of Olympic-Galata managed to qualify for playing in the top Bulgarian football division (The “A” Republican Football Group – “A” RFG) after a strong performance in the lower division during the previous season. The team caused a sensation because it was the first village team to qualify for the top football division in Bulgaria. This fact not only contributed to raising the self-confidence of the Muslims from the Teteven region, and particularly of the population of the village of Galata, but also opened-up certain employment opportunities. In accordance to the regulations of the Bulgarian Football Union, there were certain requirements for football stadiums, which host games from the top division championship. Taking into account these requirements and the increased interest towards the matches played by the team, the club sponsors invested in renovating of the Town Stadium of Teteven, changing the name of the club to FC Olympic-Teteven. As a result there were a great number of temporary job openings for construction workers - mostly young men from the village of Galata, preferably relatives or neighbours of the contractors and the players. Women were also engaged in sewing various articles and accessories used at the football games of the team.

During contemporary football derbies, especially in a crisis period, the interaction during the game usually facilitates communication among the people in different spheres. The perception of the football game by the Bulgarian Muslims from the Teteven region includes all the conditions for a satisfying leisure experience – available designated space, a limited set of rules, temporary timeframe and an understanding that the outcome of the game does not affect the actual life of the group as a whole. However, despite the stagnation caused by the crisis, there is obvious desire for good achievements and victory, which is equally strong in the players and in the supporters from the community. Thus each game becomes also a manifestation of the unity of the group. And in this particular manner sport initiates integrative moods and emotions on both local and national level[10]. For

[10] E.Hobsbawm. Mass-Producing Traditions: Europe 1870-1914. – In: E.Hobsbawm, R. Ranger (eds.). The Invention of Tradition. Cambridge, New York, 1992, p.288-289.

example, according to P.Petrov, the population of some settlements and regions in Bulgaria use wrestling as one of the source elements of their identity[11]. It can be added that the same is applicable to football. Among the Bulgarian Muslims in the Teteven region, a specific local patriotism is also observed at the level of individual settlements. Some initial hesitation whether the club should be based in the neighbouring village of Bulgarski Izvor, also populated by Bulgarian Muslims, was categorically resolved in favour of the native place of the main sponsor – the village of Galata.

Notably, during a game in Teteven in the summer of 1997, there could be observed high attendance by representatives of the community, including women, some of whom wore the traditional costume, characteristic for the group[12]. At the bus station there were special buses providing free transport for those wishing to watch the football game. *"At a match it is full, like at a fair"*. Enthusiasm and a sincere interest towards the game were evident not only on behalf of the supporters, but also on behalf of the team sponsors. The striving to win, to be superior to the others and to prove one's reputation could be described as a primordial desire, but in the contemporary context there could also be added an element of testing and demonstrating the integrity of the group. Despite the fact that modern football is rather commercialized and has lost some of its spontaneity, each game raises the spirit and solidarity of the comparatively isolated community of the Bulgarian Muslims from the Teteven region.

The football game can also be perceived as a release of the tension in the relations with the surrounding communities. The Christian population of Teteven followed and participated in the club events with certain pride, although the young people privately joked that "Mount Olympus" will probably be renamed "Galata". Great numbers of people living in the region attended regularly the games at the Teteven stadium, especially as FC Olympic-Galata defeated FC Neftohimik-Burgas - the second finalist of the top football division during the previous championship season. It could be said that the traditional stereotypes and concepts in the relations between Christians and Muslims continued to surface, but at the same time they were normalized and channeled in a positive direction.

---

[11] P.Petrov. Koi, kak, za kakvo se bori. Za niakoi upotrebi na 'narodnata borba', [Who Wrestles, How and for What? On the Utilization of 'Folk Wrestling' in Bulgaria], *BE,* 2000, No3, p.40.

[12] For greater detail about the costume of the Bulgarian Muslim women from the region of Teteven as a sign of identity, see: I.Kyurkchieva. Svetat na bulgarite myusyulmani ot Tetevensko – prehod kam modernost. Sofia, 2004. [The World of Bulgarian Muslims from the Teteven Region – Transition to Modernity]. Sofia, 2004.

The football team also became a vehicle for presenting the local population at national level. Particularly curious is one story, narrated by the mayor of Galata, about a woman from another settlement with the same name located in the distant Varna region, who expressed gratitude that the team was popularising the name of her village as well and invited the mayor to be a guest at her house.

The football games reflect the desire of the community of the Bulgarian Muslims from the Teteven region to demonstrate their unity and self-confidence, as well as the means to present the community on a more expanded scale – to the neighbouring Christian population and to the whole nation. In this context, the football game during a period of crisis could be viewed as specific reaction to the crisis, as well as a way of addressing some issues such as employment.

In conclusion, it could be summarized that the well-preserved kinship network among the Bulgarian Muslims from the Teteven region has provided a stable and well operating mechanism for preservation of the community in the contemporary setting. The period of transition presents a challenge to the solidarity of the group, but it also mobilizes it in a critical social situation. The football events provide opportunities for the manifestation of identity and practical mutual assistance, and are an example of the deeply rooted vitality and manifested solidarity of the community.

Football at the End of the Crisis[13]

In the beginning of the 21st century, the crisis began to be gradually overcome in Bulgaria. This was also true for the Rhodope Region, although it was one of the regions most affected. The revival of business provided a good foundation for the revived football enthusiasm. The improvement of people's economic situation and employment opportunities helped the rapid overcoming of the psychological impacts of the crisis. As a result, there could be observed a higher interest in entertainment and leisure activities. This also contributed to a boost in sports enthusiasm. The selection of the reviewed football clubs from the Garmen Municipality is based on their participation in the spring football season of 2007. All the teams in the municipality are playing in the Mesta Zone of the Fourth Blagoevgrad Division. The team of the village of Dabnitsa ranks 5th, the club of

[13] This part of the paper dedicated to the football revival in the Garmen Municipality was presented at the 6th Bulgarian-Serbian scientific conference "Societies, Transformations, and Cultures. The Post-Socialist Everyday Life in Bulgaria and Serbia" held on October 6-9, 2008 in Varna, Bulgaria. It is to be published in 2009 in the compendium of conference papers.

the village of Dolno Dryanovo is 7th, and of Ribnovo – 15th. In the analysis I have also included the team of the village of Debren, which was still in the process of registration, but demonstrating serious intent to participate in the regional football subgroup[14]. It should be mentioned that the ethnic and religious composition of the teams in each village corresponds to the demographic makeup of the respective villages.

About one-third of the villages in the Garmen Municipality restored their football clubs between 2004 and 2007. The initiative was usually of the young people from the villages who were supported by enthusiastic teachers, businessmen or mayors. The euphoria gradually affected most of the local people. The emphasis in stimulating the amateur football players was that they represent their settlement: *"You are representing Ribnovo, you are not playing for me, you are playing for Ribnovo!"* (a man born in 1956). This type of local patriotism was the main incentive for participation of the young players: *"We have a craving; there is something we can do for the village..."* (a man, Dabnitsa, born in 1986). In this way, through the success of the teams they raise the prestige of the village and contribute to boosting the self-confidence of their fellow villagers. Each game provides grounds for consolidation of the respective community and manifestation of its solidarity.

The organizers, sponsors and presidents of the teams are convinced that the revival of the teams is important factor in reducing alcohol abuse, hooliganism or computer game addiction among the young people in the villages. Naturally, the basic incentive for players and fans remains the pleasure of the game and the emotions brought by the sports competition. This is well summarised by the words of the president of one of the teams: *"I find pleasure in seeing 400-500 people from the village who have come out and enjoy themselves for two hours. This is enough for me. We have no other entertainment, no concerts, no cinema, and instead of watching some crazy film on the TV..."* (a man aged about 50). Everybody is aware that qualifying for a higher division is difficult and almost impossible. The aim of the amateur footballers and their fans is to participate in the events and to have fun. The mayor of the village of Debren considers that the registration of the team is *"probably the best initiative, around which more people have rallied and all have decided to work with zeal"*. The restoration of the teams is the result of the overall revitalization and the

[14] There was one more team from the municipality in the Mesta Subgroup – FC Mineral from the village of Ognyanovo, but it had ceased to take part in the championship because of the leave of the football players, recruited in the army: See *Hronika* [Chronicles] Newspaper, 21-27.03.2007, No266, p.6. It was at the bottom of the ranking with the worst performance.

changed social and economic situation in the region. The revival of the football clubs in this case can be construed as a sign of exiting the crisis period.

In Bulgaria, same as in many other parts of the world, football is traditionally related to politics. The intertwining of relations in politics and football has not bypassed the small municipality of Garmen and is part of the daily life of the people. Local authorities can provide various types of support to the development of sports. At the same time, sports can help boost the prestige of local politicians and the attainment of certain political objectives. The managers of the football clubs often think that the authorities are obliged to support their activities: *"The authorities pay no attention to sports. I care about football, but I don't get support from anyone."* This view probably reflects the typical conviction that it is the obligation of the state to have a dominant role in such undertakings. On the other hand, it indicates certain disappointment with the powers that be.

The greatest hopes for the development of football in the Municipality of Garmen rest on the cooperation of the municipal authorities represented by the Mayor of the Municipality and the Municipal Council. The municipality covers the legal fees and assists for the registration of some of the teams. The interviewees emphasized that the current municipal mayor is personally interested in the fate of the clubs and is dedicated to supporting sports. At the same time, his political opponents have used football as a way of criticizing his administration. One respondent expressed an opinion that certain funds were provided to the clubs in the municipality *"just before the elections"*, while *"no one knows where the budget surplus went"*[15]. In this way sport appears to be a major factor of strengthening the positions in local politics. The respondents did not spare criticism of some technical obstacles related to municipal support: *"The municipality gives some funds, but first you have to buy and then present an expense document in order to receive some money from them."* Generally, the approach to development of football is invariably a part of the overall image of the mayor of the Garmen Municipality.

The village mayor is the other key figure in the organization of football affairs. The mayor of the village of Dolno Dryanovo stated that *"mayor and team are related"* and this is a guarantee for the success of the undertaking. This same mayor is also a businessman and he can provide some of the prizes and the material incentives for the amateur footballers. Some people from the village have stressed that they would vote for him only if

[15] At the same time, the same respondent did not omit to note that *"he has been impeccable as a mayor"*.

he *"leads the team"*. The overcoming of the crisis in the past few years has led to positive change in the people's economic standing and some of them have accumulated certain wealth, which has also reflected on the sports activities. The mayor in the village of Debren has also been actively involved in the football initiatives, which are currently at the point of registration of the club. The involvement of mayors with the management of football clubs also has traditional roots in the socialist period: *"When I was mayor in 1985, I was elected chairman of the club and of the athletics society"* (a man, Ribnovo, born 1956).

Mayoral candidates also make use of football issues in their campaigns. In the local elections in the fall of 2007, the mayoral candidate of the Movement for Rights and Freedoms (MRF) for the village of Dabnitsa was president of the football club, a well-known businessman and, in fact, the main sponsor of the team. He had the ambition for a change: *"If I win, I shall develop projects proposals and will apply for funding.... My ambitions are not to receive a salary, but if I can, to do something during the very first year...."* The involvement with the football team was an important argument in competing with the other candidates' pre-election pledges. *"I've been trying to find support for a third consecutive year!"*

The mayor of the village of Ribnovo provided the sports outfits for the local team, which was highly appreciated by the president of the club: *"...I don't care how he did it or whether the MRF has provided it."* He is willing to ignore his political preferences, because he considers that such valuable acquisition is very important for the team.

Based on the above, it can be inferred that in most of the studied locations the mayors seek public acknowledgement through support for football. Their input for the registration, development and achievements of the teams often traces their entrance and affirmation in politics. In many parts of the world, football has been discussed as a vehicle for the social models around which a certain political behaviour is structured[16]. It is not only a reflection of relationships in society, but also part of their development, models, social and political evolution[17]. The qualities and skills of the mayors in the Garmen municipality have often been assessed in the light of their actions in support of the football clubs. During elections, some of the interviewees based their support for the respective candidate on the candidate's commitment to the football idea, despite the interviewee's personal partisan preferences. Another important factor for the success of

---

[16] V.Duke, L.Crolley. Football, Nationality and the State. New York, 1996, p.6-7.

[17] E.P.Archetti, A.G.Romero. Death and Violence in Argentinean Football. – In: R.Giulianotti, N.Bonney and M.Hepworth (eds.). Football, Violence and Social Identity, London, 1994, p.69.

sports initiatives has been the connection between politics and business. This reflects the new economic realities at the end of the crisis in Bulgaria. Some of the mayors and mayoral candidates are businessmen and most of the organizers of the teams have connections with representatives and executives of various companies.

The football matches and the accompanying emotions often reflect existing stereotypical attitudes, local ethnic and religious interrelations and rivalry between villages.

In the Garmen municipality, where Christians and Muslims have lived together for centuries and their relations became regulated with time, the matters of religion are mostly avoided in general conversations. *"Awkward topic. We mostly discuss other topics – the problems. There's no work for me, and there's no work for him. We talk about football, religion is not involved."* (a man, born 1958). The football topic proves to be the one integrating people in their everyday life. It is also a way of avoiding harsh disagreements and concentrating on common interests. However, football can also be the spark that flares up existing antagonisms. One 34-year-old male interviewee comments: *"I don't even take sides in football in order to avoid provoking conflicts"*. The football matches can be the specific occasion for surfacing of traditional disagreements, for instance, between Turks and Bulgarian Muslims, although they are not divided by religious affiliation. Another interviewee's opinion confirms the thesis that *"the differences are exacerbated during football matches"*.

An example of this were the events during a football match in the month of March, 2007, between the teams of the village of Dabnitsa[18] and of the village of Ribnovo[19], which took place in the stadium of the latter village and ended with a score of 5:3 in favour of the home team. Four men from the village of Ribnovo, aged between 20 and 70, were sentenced by the Regional Court in the town of Gotse Delchev not to attend sports events in the municipality and to pay a fine for football hooliganism. According to journalists "the accused entered the football field armed with wooden poles" and attacked the supporters of the opposing team. As the cause for their behaviour they stated that there was *"a penalty kick, which was not awarded to their team"*. Witnesses from the village of Dabnitsa

[18] Among its 2115 residents there are Turks, Bulgarian Christians, Bulgarian Muslims and Roma.

[19] The population of the village is 2753, all of them Bulgarian Muslims. The village of Ribnovo is the village located at the highest altitude and the most colourful village in the region, famous throughout the municipality for its Friday market, and in the country for the specific attire of the women and for the *gelina* (a custom, whereby the face of brides is "painted").

claim that they were pushed around, insulted and humiliated with the pejorative "filthy Turks"[20].

The viewpoint of a respondent from the village of Ribnovo was that the conduct of his fellow-villagers was *"sheer stupidity"*, which *"even came out in papers, some journalist stuff"*. The elaboration was that some of the fans from the village of Dabnitsa, headed by the team trainer, entered the restricted security area of the football field. Then some avid fans from both sides crossed the field. The regulations in such cases are clear: the Ribnovo club was banned from hosting one match, which the Ribantsi[21] accepted as justified. *"Our club was fined 55 Leva for offensive ethnic remarks, because these two old men[22] allegedly said "filthy Turks" and the like.... Out of sheer stupidity."* The inclination towards aggression and football hooliganism can be explained by the identification of the supporters of the team with a certain territory,[23] and the desire for affirmation of their collective identity. The football vernacular and the general language used by fans during football matches often reflects ethnic and social stereotypes, which are insulting to the opponents[24] and aim to display the distinctive group affiliation of the supporters.

It is also necessary to present the other viewpoint of the described events. According to the statement of a respondent from the village of Dabnitsa, the fans from the village of Ribnovo reacted so sharply because *"some of our people were speaking in Turkish and some of the locals attacked us..."* It is possible that the ensuing heated exchanges reflected traditional disputes and rivalry between the two villages. Another explanation for the reaction is the perception that the Ribantsi are more religious and have the self-confidence of an autochthonous population: *"They, the Ribantsi, want to prove that they are a local population, while we have come later[25]."* The guests of the village of Ribnovo notice that some of the women still wear the traditional headscarf, whereas *"our men cannot go there, wearing shorts. They have adopted things from Islam, but the extreme things..."*

---

[20] *Hronika* [Chronicles] Newspaper, 14-20.03.2007, No254, p. 6.

[21] "Ribantsi" is how the people of the village of Ribnovo, call themselves, as do people from neighbouring villages.

[22] Two of the culprits were aged respectively 70 and 62.

[23] E.P.Archetti, A.G.Romero. Death and Violence in Argentinean Football. – In: R.Giulianotti, N.Bonney and M.Hepworth (eds.), Football, Violence and Social Identity, London, 1994, p.42.

[24] U.Merkel. Football Identity and Youth Culture in Germany. – In: G.Armstrong and R.Giulianotti (eds.). Football Cultures and Identities, London, 1999, p.58-59.

[25] The reference is to the period of the Ottoman Empire.

According to the mayor of the neighbouring village of Dolno Dryanovo, football must convey solidarity, tolerance and mutual respect. Indeed, for most of the fans it remains just pleasant spending of leisure time, which brings joyful emotions: *"When there is a football match it is a holiday for us..."* The mayor of the village of Debren, who attended the first friendly match of their revived team, experienced unforgettable moments: *"And there were many people. The slope there is like the Amphitheatre in Plovdiv and people were perching all over. A pleasant view. When I saw the people I was really moved because this was a view that had been forgotten. When one sees people for the first time in so many years, it is very pleasing."*

Football encounters prove to be the place where traditional prejudices surface. The existing attitudes in the relations among the different ethnic and religious groups in the municipality are reflected in the heat of the sports experience. This is due to the fact that football is a way of expressing and preserving of identity.[26] And the feelings associated with the game contribute to the constructing of this identity. At the same time the emotions during the football game are a well-functioning mechanism for overcoming accumulated negative perceptions, experiencing a kind of catharsis, synchronization of the interrelations and channeling them in their traditional course. Regardless of the period, whether in a time of crisis or after it, the ethno-religious relations find expression at the football matches. This is understandable since football contributes to the formation of the perceptions related to self-identification and differentiation from the others. Nevertheless, most interviewees consider that the passion for sports supersedes all other emotions at the derby match, even in some conflict situations. The perception of sporting events as key collective rituals makes them an important catalyst, boosting the sense of community in the monotony of everyday life.[27]

Conclusion

The presented case studies are aimed at corroborating that there are specific human activities that accompany the crisis throughout its development - from its shocking start to its slow and painful end. One of these activities is football, which is something more than a game, something

[26] V.Duke, L.Crolley. Football, Nationality and the State. New York, 1996, p.6-7.
[27] G.Hare, H.Dauncey. The Coming of Age: the World Cup of France '98. – In: G.Armstrong and R.Giulianotti (eds.). Football Cultures and Identities, London, 1999, p.50.

more than passion. It raises the spirits in times of hardship and it fills the hearts with joy when the hard times slowly begin to fade away. The relation between football and politics is present in the social life not only during a time of crisis, but also after it. It is typical for both democratic and totalitarian societies. It is part of all political models and reflects the processes taking place in the social environment. Football games become arena for display of the identity of different groups. Through such events the groups make a statement about their emancipation, distinction from others and self-affirmation. Football divides, but also unites the supporters and that is one of its main characteristics, which is visible during a period of crisis. It is a game engaging people's creative passion, providing an emotional outlet during the hard moments amidst the crisis and during the emergence of the first hopes for its eventual overcoming.

References

Archetti, E.P., A.G.Romero. Death and Violence in Argentinean Football. – In: R.Giulianotti, N.Bonney and M.Hepworth (eds.), Football, Violence and Social Identity, London, 1994.

Booth D. Escaping the Past? The Cultural Turn and Language in Sport History. *Rethinking History*, Vol.8, No1, March 2004, p.103-125.

Cheska A. Sport Spectacular. A Ritual Model of Power. *International Review of Sport Sociology*, Vol.14, 1979, p.51-72.

Cholovich, I. Football, Hooligans and War. - In: N. Popov (ed.), The Road to War in Serbia, Trauma and Catharsis. Budapest: CEU, 2000.

Duke V., L.Crolley. Football, Nationality and the State. New York, 1996.

Filipova-Kyurkcyhieva, I. Kinship and Solidarity Among the Bulgarian Muslims of the Teteven Region. *EB*, 5, 2001, p.139-150.

—. Football and the Political Symbolism in Bulgaria in the 1980s and 1990s. *EB*, 10, 2006, p.345-362.

Georguieva B. Sportat v balgarskia grad (kraya na XIX – sredata na XX v.). [Sports in the Bulgarian Town, End of the 19th–Middle of the 20th Century]. – In: *Godishnik na Sofijskia Universitet* [Yearbook of the Sofia University]. Vol.93-95, 2000-2002. Sofia, 2003, p.69-118.

Hare G., H.Dauncey. The Coming of Age: the World Cup of France '98. – In: G.Armstrong and R.Giulianotti (eds.). Football Cultures and Identities, London, 1999.

Hargreaves, J. Sport, Power and Culture. A Social and Historical Analysis of Popular Sports in Britain. New York, 1986.

Hobsbawm, E. Mass-Producing Traditions: Europe 1870-1914. – In: E.Hobsbawm, R. Ranger (eds.). The Invention of Tradition. Cambridge, New York, 1992, p.263-307.

Iliycheva, M. "Faithful Until Death": Sports Fans and Nationalist Discourse in Bulgarian Internet Forums. Budapest, 2004.

Kozanoglu, C. Beyond Edirne: Football and the National Identity Crises in Turkey. – In: G.Armstrong and R.Giulianotti (eds.). Football Cultures and Identities,

Kyurkchieva, I. Svetat na balgarite myusyulmani ot Tetevensko – prehod kam London, 1999. modernost. [The World of Bulgarian Muslims from the Teteven Region – Transition to Modernity]. Sofia, 2004.

Lipsky, R. Towards a Political Theory of American Sports Symbolism. – In: Harris, J., R.Park, (eds.). Play, Games and Sports in Cultural Contexts. Champaign III, 1983, p.79-91.

Merkel U. Football Identity and Youth Culture in Germany. – In: G.Armstrong and R.Giulianotti (eds.). Football Cultures and Identities, London, 1999.

Newspaper *Hronika* 2007.

Petrov, P. Koi, kak, za kakvo se bori. Za niakoi upotrebi na "narodnata borba', [Who Wrestles, How and for What? On the Utilization of 'Folk Wrestling' in Bulgaria], *BE,* 2000, No3, p.28-47.

—. Praznichni tradicii i politicheski praktiki. Dissertation work, BAS, Sofia. [Festive Traditions and Political Practices, PhD Work, Institute of Folklore at the Bulgarian Academy of Sciences], 2002.

Wagner, E. An Overview of Sport in Asia and Africa. – In: E.Wagner (ed.) Sport in Asia and Africa. Comparative Handbook. New York, 1989, p.3-11.

Internet Sources

http://www.garmen-bg.com - Development of the Municipality of Garmen for the Period 2007-2013.

Illustrations

1. Football team of the village of Galata

Abstract

The current text presents a view of football and related activities as part of the social developments in Bulgaria during the last decade of the 20th century and the first decade of the 21st century. The social, economical and political crisis after the changes in 1989 gradually affected all aspects of everyday life. Only recently – in the past 4-5 years, there is an emerging public sentiment that the crisis is being slowly overcome. The research presents two separate examples of football related activities and emotional outbursts among the supporters of two village football clubs. They are studied as part of the events resulting from the overall crisis.

The relation between football and politics is present in the social life during a time of crisis. It is part of all political models and reflects the processes in the society. Football games become arena for display of the identity of different groups. Through such events the groups make a statement about their emancipation, distinction from others and self-affirmation. Football divides, but also unites the supporters and that is one of its main characteristics. It is a game engaging people's creative passion, providing an emotional outlet during the hard moments amidst the crisis and during the emergence of the first hopes for its eventual overcoming.

Iva Kyurkchieva, PhD in ethnology, is a junior associate research fellow at the Ethnographic Institute and Museum - Bulgarian Academy of Sciences. Her major interests are in the area of the interrelations between the ethnic groups in Bulgaria, the ethnic processes in our time and anthropology of sport.

An Analysis of Western Chinese Minority Youths' Susceptibility to AIDS

MINGXIN LIU AND YUANGAO HOU
ASSOC.PROFESSORS, SCHOOL OF ETHNOLOGY AND SOCIOLOGY, MINZU UNIVERSITY OF CHINA

Since 1985, when AIDS was first discovered in China, the disease has quickly penetrated and exceedingly engulfed the Chinese population. The number of those infected with the virus has rapidly increased. Among those in Western China's minority communities, the speed of propagation of AIDS has been especially high. The National Report of those infected with AIDS states that Yunnan Province has the highest number of AIDS patients, followed by Xinjiang, Guangxi, Guangdong, and Sichuan respectively. From these regions, we can see that the population of minority nationalities with AIDS is proportionally exceeding the infected Han nationality population. The entire population of minority nationalities in China accounts for 8.46% of the whole population currently, while ethnic people made up 36% of the registered people who tested positive to HIV/AIDS in 2001 (National Health Department, 2001). The minority groups most seriously and critically facing the AIDS epidemic are the Jingpo, Yi, Achang, Dai, Hani, Lahu, Uyghur, and the Hui.

The Chinese government has continuously paid close attention to controlling the AIDS problem, adopting various effective methods for implementation, announcing publicly the "four remittances and one care "policy (free anti-virus drugs for rural and urban residents with financial difficulties, free blood testing and voluntary information service, free AIDS interdiction of mother-infant transmission, free education fees for those orphans with infected parents, taking care of HIV patients and their family), promoting financing support for government programs to enhance the prevention of AIDS. However, within the process implementation there are specific things left undone, for example, focusing on publicizing within the mainstream media but disregarding the audience's capacity of acceptance and education level; focusing on health education towards teenagers in school, but not acknowledging how to efficiently afford guidance to those not attending school. In fact, ethnic minority teenagers in west China have a rather high drop-out rate, far above the average level. Due to the lack of opportunities to receive education, as well as their low level of Mandarin comprehension, ethnic minority teenage youths are faced with narrow channels through which to obtain information, leaving them with relatively poor basic health knowledge. The poor organization of traditional communities of minority groups leads to leniency of sexual behavior among minority youths, which causes the youths' greater involvement in high-risk conduct. According to a survey, drug addicts between the age of 17 to 36 make up 65% of drug addicts in Yunnan province. In some areas it has reached 75%.[1] For these reasons, one of the main foci of the current AIDS prevention programs is to enhance research on the general characteristics of western minority youths and take effective measures to promote their ability to prevent and control the AIDS epidemic.

1. Basic Characteristics of Western Minority Out-of-School Youths

In general, western Chinese minority youths have the same basic characteristics as their peers, while they also have their own particularities. These particularities are outlined as follows:

1) In the psychological domain, minority youths appear to have high curiosity and vanity. They enjoy comparing themselves with others. For youths, it is a general psychological characteristic to be curious of the outside world, and this is also a common characteristic of minority youths. According to a survey, many teenagers use drugs to fulfill their curiosity. Drug use is a way to "try out a 'comfortable' feeling that is described by others", and often leads to addiction. At the same time, the factor of most out-of-school youths generally comparing themselves with others is the main reason why they choose to use drugs and have multiple sexual partners. Because of the fact that drugs are an expensive commodity, and many people cannot afford them, using drugs has become a sign of "I can get something special that others can't" in the eyes of these youths. When it comes to the question of relationships and sexual behavior, it is not only males but also females who take the fact of being in a relationship with

[1] Jia Zhongyi. A Report of the Policies Towards Controlling of Drug Influence and Practical Investigation of Yunnan Provice. *Journal of Investigations of Problems of Drugs and AIDS in Ethnic Minority Areas*, Hou Yuangao, (ed.) March 2005, p.9.

multiple partners and gaining sexual experience as a way to show off to their peers.

2) In the context of social intercourse, out-of-school youths can be understood by addressing three factors: gender, age, and occupation. Members of their social circle are determined by gender, age, and occupation or if they work in the same geographical area. Among those social circles, faction character is clear and the concept of "Jianghu"[2] is popular. In addition, out-of-school youths who migrate to cities have great difficulties in developing intimate personal relationships with local people. This lack of relationship is due to the gap between out-of-school youths and city dwellers in communication, habits, and social status. Their communication with people within their ethnic group or people from their hometown is, therefore, enhanced. Furthermore, this leads to out-of-school youths becoming the "blind spot" of the government and community service centers and related departments. As a "blind spot" it is not only difficult for them to receive appropriate education and sufficient support from society, but they also lack motivation to gain related information on their own, making the channel for information and opportunities even narrower. This is ultimately how out-of-school youths become candidates or victims of high-risk behavior.

3) In regards to marriage, relationships, and sex, out-of-school youths generally appear to have an open idea towards pre-marital sex, and the concept of being in a relationship and getting married at an early age is popular. However, they lack the consciousness of having safe sexual intercourse. Their first sexual experiences occur at an early age, often with multiple partners. Because the social environment of minority areas is facing rapid development and opening up, minority youths are now hunting for jobs outside of their hometown. In addition, the opening up brings pornography and promiscuous videos and books into rural communities. Traditional conceptions of sexuality are thus changing and youths are gaining more sexual experience before marriage. In Dehong, Yunnan Province, for example, the Jingpo minority youth are free to have pre-marital sex, and most of the males' first sexual experiences occur on average before the age of 15. By the time they reach the age of 20 some youths have had 15 to 20 sexual partners.[3] Women are free to have pre-marital sex as well, but their sex life after marriage is often stable and

[2] "Jianghu" originally means everywhere. Here means activities' non-routine. People uphold personal loyalty in "Jianghu".

[3] Study Centre for Western Development in MUC. Research Report on Risky Behaviors of Out-of-School Ethnic Children and Young People and Their Vulnerability to HIV/AIDS in Five Western Provinces, 2006, p.21.

monogamous. After marriage, males are allowed to date younger women in the village, and uphold the saying "your wife is your own, girls are for everyone," which is popular among the villagers. This reflects some of the characteristics of relationships within the Jingpo ethnic group. The Dai and Hani ethnic groups are also more tolerant towards pre-marital relationships and sex. In these ethnic regions, marriage and relationships at an early age are normal, especially in rural areas. According to a survey, the percentage of youths between the age of 15 and 19 who have not reached the legal marriage age yet live together or get married in the western China has reached 2.6%, double the proportion of their peers who are married (1.3%). Of this group, the proportion of females getting married at an early age is 4.2%, which is three times that of males.[4] Young women getting married and having children too early is bad for the mother physically and psychologically, and also difficult to supply their children with better education, increasing the pressure of bringing up a child. In a survey of out-of-school youths in Yunnan province, the earliest case of first time sexual experience was at age 12. Among our 69 interviewees who responded about their numbers of sexual partners, the number is from 1 to 23. On average, it is 3.04 people. People who have sexual experiences but who do not use a condom make up 61.5%.[5] The fact that they have multiple sex partners and lack awareness of safe sex has made the western minority out-of-school youths more susceptible to the risk of STDs.

4) In the rural areas, out-of-school youths have few choices in terms of entertainment and leisure activities, usually chatting with friends, drinking, singing, listening to music, and/or playing billiards. In urban centers, out-of-school youths have various ways to entertain themselves, such as going to discos, nightclubs, mahjong houses, tea houses, pool halls, internet bars, and ice skating houses. They live a life of no regularity, sleeping in the day and going out at night, and many of them have developed habits of gambling. Male youths usual go to internet cafes, bars, and electric games houses, while female youths usually go to pubs and karaoke bars. Being in an environment where they are smoking, drinking, dancing, and singing, they are more prone to sexual immorality and easily driven towards drug consumption.

5) The employment rate of western suburban out-of-school youths is comparatively low, with unstable employment, making them in great need

---

[4] An Guoqi, Deng Xiquan. Report on China's Western Youths Development, *China Youth Research,* 2007, No1, p.25.

[5] Ding E. A Comparative Study of AIDS Susceptibility of Three Ethnic Minority Groups Out-of-School Youth High-Risk Behavior, Lancang, Yunnan, Ethnic Minority Youths and AIDS Senior Research Meeting, Dec. 2005, p.37.

of welfare and security. According to a survey, the rate of underage youths who are "not in the labor market" are comparatively high, especially those between the ages of 15 to 19, which was found to be 55%.[6] Youths of this age group have difficulty obtaining employment contracts, leading to a lack of pay or overdue salaries. Most of these out-of-school rural youths move to big cities, go abroad in search for employment, yet many remain unemployed with nothing to do at home. Of those who go to cities in search for jobs, many can only find blue-collar work, or "heavy work" – dirty, tiring, bitter, and risky because they don't have residential city permits. Young women usually find jobs in the service industry, such as hotels, hair salons, and massage houses. Their salaries are extremely low and living conditions are terrible, and even their right to go on break is not protected. Those youths who stay in rural areas are mainly involved in agriculture. However, they often lack necessary agricultural skills and training, as well as beneficial information on the agricultural market. These have become their major survival and development problems to cope with.

2. Analysis of Reasons for Western Ethnic Minority Out-of-School Youths' Susceptibility to AIDS

2.1. Economic Aspects

The main reasons why western minority out of school youths are involved in high-risk conduct is the sluggish development of society and continued poverty. Those rural out-of-school youths who are struggling to change their situation of being raised in less developed areas often choose to go to find jobs in cities. However, speaking with a thick accent and not being well educated creates difficulty in communication and their lack of skills prevents these youths from finding a city job. In addition, their unstable income and their need for welfare leads to riskier behavior such as being engaged in illegal conduct, like stealing, robbery, drug trafficking, prostitution, and the abduction of people. These high-risk illegal activities tend to make youths psychologically highly stressed and nervous for long periods of time. To release this stress and also to gain an unattainable sense of happiness and a "high" in real life, using drugs and going whoring have been the most direct way of escape. Once youths are addicted, money to

---

6 An Guoqi, Deng Xiquan. Report on China's Western Youths Development. *China Youth Research*, 2007, No1, p.28-29.

buy drugs becomes the main financial pressure. In fact, it is less financially stressful to take drugs via injection (mainline), bringing users to higher level and the sharing of needles becomes part of a destined outcome. Thus these youths fall deeper into a worse circle of poverty and their high-risk conduct. When a woman has nothing, her body becomes her only capital. Women sell their bodies to live, and are often unable to control the use, or lack of use, of condoms. The extremely low rate of condom use has made prostitutes the highest risk community for AIDS. Prostitutes are spreading AIDS to a larger population through their job, and even spread AIDS to their hometown communities when they return home from the cities.

In China, the areas with the highest AIDS infection rates, except for Henan province, are in western China. The areas most seriously invaded by this epidemic are usually rural and poor communities, and include powerless groups such as migrant workers in the cities and underpaid city residents. The examples of Dehong Dai and Jingpo Autonomous Regions, Hani and Yi Autonomous Regions of Lingcang, Simao, and Honghe, Yi Autonomous Regions of Liangshan, Sichuan Province, Baise region in Guangxi, and the city of Yining in Xinjiang prove that AIDS and poverty are inseparable.

2.2. Cultural Aspects

Traditional ethnic customs and habits have also influenced the fragility of minority youths when encountering drugs and AIDS. A tradition of sharing is an important institution in hunting and gathering and agricultural societies. Sharing is a way for the social members of a community to help and benefit each other. This institution is an integral sociocultural custom of many ethnic groups. This sharing tradition involves the communal sharing of the hunted, gifts, fancy new goods, food and drink, new experiences, and in some cases personal property. In the Yi community in Liangshan, Sichuan Province, friends and relatives are highly respected, and they uphold the conception of "if we have wine, everyone drinks it; if we have smoke, everyone smokes it." Throughout this Yi community's history only those of high status can smoke opium, making opium a precious gift on social occasions. After the influx of modern drugs into the region, the local people, without sufficient knowledge of the damage of these drugs, believe them to be precious possessions, and to be shared among friends. This, in fact, led to the usage and spread of drugs in the area. In Uyghur societies in Xinjiang, northwest China, people respect friendship and are also willing to share with friends. In the past, marijuana was not considered a drug, and sharing marijuana was a basic gesture of hospitality. After marijuana was replaced by heroin with the influx of modern drugs into the region, more social damage has taken place in the form of STDs and AIDS. Too often, people don't pay enough attention to the association between heroin and STDs or AIDS. Sharing needles during heroin use is still a common phenomenon in these local areas.

2.3. Education Aspects

Due to the undeveloped economy, many families in the western minority areas are struggling at the poverty line. In fact, many ethnic minority families have to force their children to drop out of school because of their inability to afford educational fees. In the National Education Report of 2005, about 8% of the whole country has not achieved 9 years of compulsory education. The areas lacking 9-year compulsory education are mainly located in the west. The outcome of a low education level has resulted in children's low ability to differentiate between things, and their inability to reject interfering, harmful situations and activities. Ultimately, their chance to get hurt is greatly enhanced.

Education plays an important part in the fate of these ethnic minority youths. Currently, in the process of the collective editing of textbooks, ethnic and regional characteristics are left out, ignoring the requests of rural area educators. This has resulted in a state agency silencing of ethnic minority groups in the educational institution. Collective edited textbooks have not implemented the use of local minority languages resulting in many hardships and difficulties for minority students to learn whose mother tongue is their minority language. In effect, this depresses students' desire to learn.

The huge gap between the outcome and the prospects of education has made parents and children consider schooling an unnecessary investment. The ideal value states that if only you receive more education, better opportunities to obtain a job will be available, and the quality of your life can be advanced and your social status can be promoted. However, in reality, what we have seen, particularly among female students, after overcoming all the odds to finish middle or high school, they still lack the opportunity to obtain further education or a job. They have to go back to the fields and conduct housework. They feel there is simply no way to achieve their goal of changing their social identity and living conditions. This sad reality directly influences parents' and children's attitudes towards education and their passion for continued schooling. Eventually a new opinion towards education is spread of "studying is of no use" and "knowledge is cheap".

According to a survey in Liangshan, Sichuan Province, in the past 2 to 3 years the village primary school of Butuo county has exempted all schooling fees. The Temuli central primary school also reduced its educational fees to only 1/4 of the original amount by September 2005, which is about 36 yuan RMB per year per student. Although this has started to solve the economic problem of student enrollment, the rate of enrollment for girls has not improved. This has much to do with the spreading of opinions mentioned above. Most youths in rural areas in the west have only received a primary school education, and only a handful will go on to receive a higher education.

The problems within the village schools have consequently resulted in the problems of the out-of-school youths. According to a survey, rural communities are short of educational resources, the administration is weak, the "Pu Jiu" (9-year compulsory education) policy has not achieved great success, and drop-out rates are high. This situation is particularly serious in far away ethnic minority areas. The education system is oriented toward entering a higher school. This orientation is embodied in the exam system where grades outweigh moral and ethical cultivation, the sciences outweigh the arts, and knowledge from textbooks outweighs practical daily life skills. Education is focused on urban knowledge rather than rural knowledge, which is ignored. Students are often left with a lack of practical daily knowledge, and personal and ethical cultivation. In addition, some schools and teachers discriminate and exclude, expel or even request students to leave school if they are naughty and have poor grades. Instead of imposing specific educational strategies towards these troubled students, schools employ these methods mentioned above for the sake of a higher graduation rate and excellence rate, avoiding affecting school credibility. These methods consequently hurt the students' pride, creating student discouragement, depression, and even passivity and pessimism, where it is becomes easier to incite outwardly extreme behaviors. Many problems within the school system have made a mass number of students voluntarily or involuntarily leave school and become part of the out-of-school youth circle.

In regards to health education, such courses are only taught in secondary school. (In comparison, sex education in France starts at age 9. By the time students reach 13 years old, sex education has entered a compulsory education project). Because the rate of ethnic minority youths receiving secondary education is low, most of these youths lack basic health education knowledge and guidance. Also, those children in rural areas entering school at a later age have already begun puberty, making it sometimes too late. Lacking related educational guidance as well as the loose administration of some rural schools has made these innocent and excited children susceptible to unsafe sex.

2.4. Environmental aspects

The family is the first institution for children to be socialized. According to a survey, due to divorces and the absence of parents, children who are raised in an incomplete family, do not receive sufficient living resources, making them engage in arduous labor and having to cope with economic stress. Youths are driven to live in an environment where there exists little family discipline making them more susceptible to high-risk behavior and the threat of AIDS. Children who grow up in a family where family members use drug and/or have experience of drug trafficking lose their fear towards drugs. This direct influence sometimes leads them on the path of drugs use at an early age. At the same time, because the parents are not well educated they usually use very simple or violent ways to educate their children. This not only has disastrous emotional impacts, but also makes children grow up with a rebellious attitude - hateful towards family members and often trying to get revenge on society. Some become gangsters and commit crimes.

The second influence on youths is the role of the community. In some ethnic minority areas in the west, such as Sichuan, Yunnan, and Xinjiang, there has been a long tradition of smoking opium. Under the influence of such cultural traditions, young people do not have sufficient information about the dangers of this drug, and, moreover, lack a clear understanding of new types of drugs coming into the area, such as heroin. At the same time, the influence from their peers is also unavoidable. Risky behaviors spread among teenagers and youths in direct relation to the spreading of drugs. Since sharing and following others is an important aspect in getting recognition from their peers, youths easily make wrong choices when they are faced with collective pressure. Their friends usually seduce them into a certain lifestyle, such as drug addiction or prostitution, which leads them along a path of self-destruction. When these individuals have decided to correct their mistakes, they often fall back into the deception of their friends. The influence from peers has become one of the main reasons that out-of-school youths step along the path of drug use.

In minority areas in the west, guidance, education, and administrative support of out-of-school youths is insufficient in the community. Facilities for public activities are seriously needed in village communities. In addition, young people don't have many entertainment options so that their idleness cannot be altered. Meanwhile, there is insufficient direct guidance among city youths when it comes to entertainment, attracted mainly to discos, bars, karaoke halls, internet bars, pool halls, and arcades. Because of this situation it is easy for them to fall into way of crime under the influence of harmful social elements from the community and the internet.

3. Countermeasures and Suggestions

According to the analysis above, the spread of AIDS in western China and the causes that shape out-of-school youths engagement in high-risk behavior are complicated. The general situation is not positive. Based on our own research and working experience we have the following suggestions to create a better environment for youths to grow up in:

1) Governmental Stage: The government should improve legislation, establish anti-drug statutes and support legal regulation as soon as possible. At the same time, the government needs to improve and promote financing for drug control and AIDS prevention to make sure the related projects can be carried out. On the arrangement and distribution of anti-drug and AIDS prevention resources, the government should consider the current social development level of the west. The government should not only have beneficial policies for countering drug use and the spread of AIDS, but also offer support to the people in the west to improve their living standards, job opportunities, and quality and standards of education. Only by doing this, can the crisis of AIDS and drugs in the West be alleviated.

2) Community Stage: To perfect the reception of community petitions and to create a safe environment for the youths to live and interact, community control must be enhanced and every "death corner" (dangerous and risky locations) in the community must be purged. Right now the control and safety of the social environment mainly relies on the government. The district anti-drug office and the public security authority, in association with the local public security office and the village committee, are attacking the circle of drug dealers and any drug trafficking behavior hoping to cut out the supply of drugs and block the transport and sales network. At the same time, local authorities are conducting surprise investigations at hair salons, karaoke halls, disco bars, and internet cafes, to control group gambling, drug dealing, erotic entertainment services and drug use, in the purpose of protecting the residents and youths of the community. For those out-of-school youths and high-risk groups, we need to provide them with a certain degree of humanitarian support, decrease social discrimination, and create a better environment for them to grow up in.

3) Educational Stage: Efforts need to be made to enhance health education and publicity of health education towards students in school, to support community education of out-of-school youths, and to rid all "blind spots" in the education system. For the publicity of health knowledge and sex education, the PLA (Participant Learning and Activity) mode of study should be adopted, to achieve enriched human rights of youths, allowing them a fresh start in taking responsibility towards family, society, and themselves.

4) Family Stage: The main living arena of out-of-school youths is the family. Parents have the highest authority in the family. Therefore, parents should fully acknowledge the complexity of the circumstances in which their children live and promote family involvement in preventing their children's engagement in high-risk behavior. Some ethnic minorities in Yunnan province, such as Dai, Bulang, and Hani people in Daluo town of Xishuangbanna, have demonstrated the importance of family. These three minority group have paid close attention to providing their children with daily life knowledge, useful skills and various kinds of social etiquette, as well as recognizing that ethical education is one of the most important family education contents. These three ethnic groups' customs of honoring elders, taking care of the youth, and helping each other, the prohibiting of encroachment on others' property, appreciation of public order and ethics, and sustaining the purity of self by leading a clean and honest life are all emphasized through example by the honorable elders and social gatherings such as weddings, funerals, festivals, ancestor worship. It is from these social activities that they pass on their great traditions to future generations. In surrounding minority communities in Daluo town, malicious criminal conduct, such as drug dealing, robbery, rape and prostitution are also rare. From this we can see that calling for greater roles of family education is very helpful in raising youth.

5) Peer Groups: Peer groups are another domain that out-of-school youths belong to. Most of these out-of-school youths have friends with the same interests. They value friendship. They share things with their friends rather than their parents, especially topics concerning relationships, marriage, and sex, which "secret" information is only shared among friends. The head of the peer group has an overwhelming influence on every member of the group. They are privileged with a high reputation. Their conduct, thus, has a direct guiding influence on the group members. Therefore to proceed in improving education, it is necessary to first educate the head of the peer group, and use their influence to pass on the right message to other youths. Information sharing within peer groups has become an important channel for out-of-school youths to gain knowledge and related information on AIDS and drugs.

6) Non-governmental and Religious Authorities: In western minority areas we need to make use of traditional ethnic culture, to help our youth resist the invasion of drugs and AIDS. Important people in civil society, religious leaders, community committees and social organizations have a significant role in controlling drugs and the spread of AIDS. The Jiazhi (one kind of patrilineal kindred organization existed in Liangshan, Sichuan province), for example, is a basic social organization of the Yi people of Liangshan. The Jiazhi has its own laws and social etiquette, which are of great significance in civil society. In the late 1990s, when drug use began to spread in Liangshan, some Jiazhi called for social meetings to control drug use. They gathered those family members who were known to use drugs and forced them to quit, created terms of punishment for drug dealers and made it part of their community law. Some Jiazhi organized a Jiazhi patrol. If someone is found engaging in drug dealing, his/her property is taken away and that person is banished from the village. If someone is found using drugs, they are put under close observation, their social activities are restricted, and their food allotment is put under the management of the Jiazhi in order to prevent them from selling their food to buy drugs. In serious cases, some drug users could be banished from the village. Being banished by the Jiazhi means that the social welfare of the individual is lost, which for Yi people is one of most serious punishments. These solutions have worked effectively and drug dealing and drug use have been clearly reduced.[7] Jiazhi customary law is still a very powerful, effective, and restriction force in the Yi community. The Jiazhi meeting and sacrificial ceremony to build vows of promise has been successfully adopted to intervene in instances of teenagers' harmful conduct.

The influence of religion in ethnic minority areas is universal. The positive aspects of religious teachings in dealing with drug and AIDS prevention are very impressive. The Dai and Bulang people are two groups who believe in Shangzuobu Buddhism from SE Asia. Almost every Dai and Bulang village has a Buddhist temple to serve as the center of religious activity and social gatherings. Buddhism advocates its moral principles, regulations on conduct, and taboos to its believers through religious activities and education. In recent years, Yunnan provincial government organizations have copied an experience from Thailand to bring in Buddhism as an AIDS prevention method for AIDS patients' care and recovery and education for drug prevention. With the help of the religious

---

[7] Study Centre for Western Development in MUC. Research Report on Risky Behaviors of Out-of-School Ethnic Children and Young People and Their Vulnerability to HIV/AIDS in Five Western Provinces, 2006, p.46.

community, this project has achieved success in drug and AIDS control in Ruili, Dehong County.

According to a survey,[8] Mandarin language ability of out-of-school youths is extremely low, with only 32.8% fluent in Mandarin. This statistic shows that publicity in local minority languages is essential. While conducting publicity, we also have to consider people's actual circumstances of different communities and their ability to accept information, keeping in consideration that practical methods for knowledge dissemination could utilize popular means, for examples ethnic singing and dancing, theater, talk shows.

7) Out-of-School Youths: In promoting the education of anti-drug use and prevention of the spread of AIDS, we should value the subjective role of out-of-school youths, making them not only the party to be educated, but also positive participants. Out-of-school youths' characteristic as a floating population provides them with a wide range of social encounters, therefore, we should make extra efforts to make them volunteers, changing the one-directional information sharing, and to explore a more suitable public model to increase the influence of AIDS prevention and anti-drug education. According to a survey, those youths with a more stable job and skills are less likely to be engaged in high-risk behavior. Therefore, the government should focus its attention on the professional training skills of out-of-school youths, their difficulty of job hunting, guiding them to change their views of a life career, and promoting youth's ability to make a living.

In conclusion, the high-risk behavior of China's western ethnic minority out-of-school youths is interwoven with social, historical, and ethnic cultural factors. Only by mobilizing the strength of the whole society can we form a united force to prevent and control the spread of AIDS and create a better environment for our younger generations to live. In the meantime, we should be aware that AIDS prevention is not only a medical or biological problem, but also a sociocultural problem. Only when we focus on both biological and social aspects, devising powerful resolutions, can we effectively control the AIDS problem in China, and thus thoroughly handle the issue of China's western ethnic minority out-of-school youths' susceptibility to AIDS.

---

[8] The survey was carried out by Study Centre for Western Development in MUC in 2006. During the survey, 2150 questionnaires were finished in 20 target communities in five western provinces.

References

An Guo Qi, Deng Xi Quan. Report on China's Western Youths Development, *China Youth Research*, 2007, No1.

Ding E. A Comparative Study of AIDS Susceptibility of 3 Ethnic Minority Groups Out-of-School Youth High-Risk Behavior, Lancang, Yunnan. Ethnic Minority Youths and AIDS Senior Research Meeting. Dec. 2005.

Jia Zhong Yi. A report of the policies towards controlling of drug influence and practical investigation of Yunnan Provice. *Journal of Investigations of Problems of Drugs and AIDS in Ethnic Minority Areas*, Hou Yuan Gao, (ed.). March 2005.

Study Centre for Western Development in MUC. Research Report on Risky Behaviors of Out-of-School Ethnic Children and Young People and Their Vulnerability to HIV/AIDS in Five Western Provinces, 2006.

Abstract

This paper aims to study risky behaviors of young people (especially those who are out-of-school) and their vulnerability to HIV/AIDS, and analyze effect of drugs and AIDS on young people (under 18 years old) and the problems they face; analyze various factors that lead to their involving in risky behaviors; put forward policy suggestions and action strategy, and provide basis for exploring relevant policies and arranging relevant system. The young people who live in the western ethnic areas are increasingly engaging in high-risk behavior. The main reason for this is that the length of education in the western countryside is generally very short. Many children and young people are forced to cease their study during or after their primary school education and they often enter a risky environment. These out-of-school children and young people lack skills to make a living and the information needed to make wise choices about their behavior.

At present, the policy of "four remittances and one care" is gradually being implemented. But in remote and poor western regions especially in ethnic areas, the effects of drugs and HIV/AIDS on children and young people fail to catch enough attention of the whole society. Their vulnerability to drugs, HIV/AIDS and sexual diseases and the related cause analysis remains in need of systematical investigation and scientific interpretation.

Liu Mingxin, Ph.D., Associate professor of anthropology at the School of Ethnology and Sociology of the Minzu University of China (MUC), Beijing; teaching and research at the same school of MUC. Her scholarly interests are focused on the culture of Northern ethnic groups in China, shamanism and folk beliefs.

Hou Yuangao, Lecturer of anthropology at the School of Ethnology and Sociology of the Minzu University of China (MUC), Beijing; teaching and research at the same school of MUC. His scholarly interests are focused on the culture of Southern ethnic groups in China, applied anthropology and study of HIV/AIDS prevention of minorities in China.

Two Cultures' Reaction to a Pandemic (Ethnological Observations on China Combating the Atypical Pneumonia Virus – SARS, in 2003, and Bulgaria Facing a Bird Flu in 2006)

Elya Tzaneva

Assoc.Professor, Ethnographic Institute and Museum, Bulgarian Academy of Sciences

The past few years have established a new field of sweeping anthropological investigations which set as their task to collect and study the numerous examples of the ways in which crises, disasters and extreme situations manifest themselves in people's lives. This new field of study involves also theorizing on them, building possible models of their genesis, development, prevention, reaction to and consequences of, as well as showing effective mechanisms developed in human history and culture which have to be studied, improved and encouraged with the purpose of guaranteeing survival of the world. Most frequent crises are those of natural or technological character and their study from a historical and anthropological point of view concerns the extent to which man can combat them and overcome their consequences. Natural and social crises, however, featuring among which are also the epidemics, have their explanations in the understandable parameters of their geneses, fitting in suitable social niches, growth and fading out. These parameters are part of the history, culture and mentality of a given society and their study, aimed at building up protective mechanisms, is a serious and topical challenge to the humanitarian and social scientists.

A number of disciplines are already actively being involved in the interpretation of critical situations. It is their purpose to explore the possible risks, analyze the circumstances surrounding the arising crisis, and forecast it so as to reduce the elements of surprise and drama to their possible minimum. Apparently, anthropologists also have a legitimate place among the researcher of crises, at least insofar as they study the psychological, cultural and historical prerequisites for the arising and overcoming of a crisis. The place of the specialists in the wide range of anthropological studies, and especially of the ethnologists, is determined by the mechanism of the development of a crisis: more often a crisis strikes suddenly, in an unforeseeable and insurmountable way and often ruins the standardized and stereotyped pattern of interrelations between the individuals and their natural and social environment, producing misbalance. The disastrous natural phenomena – such as earthquakes, floods, landslides, tsunami waves and hurricanes – as well as those with a social background – such as epidemic diseases, the aftereffects of wars, revolutions, migrations, etc. – wedge into the well-balanced personal and group cycle of traditional culture and disrupt its normal rhythm affecting both the individual and the entire collective.

It has been generally acknowledged that the Bulgarians react to a crisis only after it has broken out; this is a sad conclusion concerning both the recent fairly hard shocks which befell our life, health, lifestyle and optimism, the political and social problems created by ourselves mingling with natural disasters such as the droughts, heavy snowfalls, floods and viruses. Bearing in mind the difficulties we have in mobilizing our efforts to overcome them, it is easy to explain why the attention has not yet been focused on what aspects of society's cultural and historical model have made it possible for the crises to break out on such a large-scale. For Bulgarian ethnology this is a whole new field of research which is still undeveloped and will undoubtedly find its eager students[1]. The reaction of the ethnologists to such critical events elsewhere in the world has been more prompt and adequate: the new trend of "Anthropology of Risk and Crises" has been in existence for more than a decade now, the researchers collecting and arranging facts, classifying and analyzing them with the purpose to interpret them within the context of cultural history and theory. And a thematic circle studying the risk factors in society which generate crises has been taking shape in the works of specialists of various orientations in the social sciences. The risks, both traditional and new, ensue from the globalization of the world, and the world economy, from the cyber war and the emergence of the genetic technologies; they are connected with the climatic changes, or the dangers presented by Al Qaeda activists or by diseases such as the 'mad cow' – all of them becoming increasingly universal. New risks keep emerging all the time connected with changes in

---

[1] The book by the Assoc. Prof. N.Kolev Kritichni situatsii i narodniya mirogled u balgarite [Critical Situations and the Popular World View of the Bulgarians]. Veliko Tarnovo, 2003, has so far been the only study giving any place to the ethnological science in the researched context.

the natural conditions, in technology, biology, not to mention the social tensions and policy-making.

Using the method of comparative analysis, in the following pages I would like to share some ideas concerning two similar social crises, which were acute in 2002-2003 in Southeast Asia and in 2005-2006 in some Balkan regions. They caused (especially the first one) sorrowful statistics, resounding threateningly for months on and throughout the whole world, viz. the atypical pneumonia syndrome (Severe Acute Respiratory Syndrome SARS) and the Bird Flu. Specialized ethnological observations and discussions could add a valuable perspective to the emergence, spread and culmination of any such crisis. In both cases mentioned the analyses sought explanations for problems such as: what features in traditional (popular) culture had stimulated the spread of the diseases; how the specific features of the political life and practice had hampered its prompt elimination, but also how they had helped subsequently; what was thought of the disease at the time – in the days of its peak – and how it is regarded now, in the process of its slowly becoming "history". In the final count, how successfully these observations have influenced the newly emerging area of research.

I am presenting a modest attempt at an answer, within the above-mentioned context, taking advantage of my being a direct observer and investigator in both situations, and, in the case of SARS, a specialist who had the chance to experience the object of one's interest and live through it. But first let's turn to the statistical data:

Table 1. Statistics of the WHO about the SARS damage provided on July 11, 2003, the last date, on which the WHO announced that information

State	Number of registered cases	Number of deaths	Number of patients
China	5327	348	4941
Hong Kong	1755	299	1433
Taiwan	307	47	***
Canada	250	38	194
Singapore	206	32	172
USA	71	0	67

Vietnam	63	5	58
Philippines	14	2	12
Germany	10	0	9
Mongolia	9	0	9
Thailand	9	2	7
France	7	1	6
Malaysia	5	2	3

Following further are the countries with no lethal cases: 4 cases each were registered in Italy and Britain; 3 cases each in India, South Korea and Sweden; 2 cases in Indonesia; 1 registered and cured case each in Switzerland, Spain, South Africa, Russia, Romania, Ireland, New Zealand, Kuwait, Finland, Colombia, and Macao. The overall statistics gives the figure of 8069 cases of the disease worldwide, of which 775 deaths and 7452 cured. The first case of the disease was registered in November 2002 and the last – in June 2003. The death rate among the patients was calculated at 10%, the highest being in China[2].

The fact is that over the long years through which the destiny of the state and the people of China were shaped by the traditional Chinese culture it is marked by the stresses of upsurge and decline. The Chinese peo-

[2] Following is a brief account of the facts concerning the crisis: before February 2003 the official position of the Chinese party and government was to cover up the situation, to deny the epidemic and not give publicity internationally to the problem. In mid-February this situation changed at a dramatic press conference of the Government, and shortly afterwards the first victim of "the policy of silence" fell: the health minister and the mayor of Beijing. Although the break was a fact and the WHO was officially informed, the information being accompanied by an apology on the part of the Chinese government, the inertia of silence was surmounted slowly. At the end of April, by the unthinkably brave act of the Chinese physician Dr Jiang Yanyong, who, in a self-sacrificial act, stated that it was only in the hospital he worked in Beijing that there were more SARS patients than the number provided by official statistics for the whole of China, matters changed drastically and the state, together with the party, together with the army and with the entire available apparatus of authority, set out on a hard and uncompromising battle against the epidemic.

ple suffered a multitude of natural anomalies; they faced, fought against and dealt almost daily with social and technological cataclysms. However, the surprise, horror and consternation at the scale of the SARS epidemic five years ago were particularly dramatic and instructive. In extreme situations (and so was the case of SARS epidemic) the patriarchal personality, both in traditional and in modern day Chinese culture, invariably turns for strength to the cultural experience of the past, struggling to survive and outlive the danger which threatens human life, health and overall existence. On the other hand, it has turned out – through the eyes of the specialist – that precisely the said cultural experience accounts for the circumstances which made it possible for the crisis to arise. The way of thinking in this case is along the line risk-crisis-culture, and notwithstanding the fact that this line was drawn later, the knowledge deduced has fed the main effort of such investigations: to find and restrict the risk fields for society, through the purpose-oriented analysis of the cultural experience and acts, which would help to avoid or reduce the effect of the critical situation.

The atypical pneumonia syndrome – severe acute respiratory syndrome (SARS) first struck the South of China – the wonderful places around the Southern Sea, which were the historical "gateway" to this part of the world and which have become an emblem of the success and the flowering of the state. The weeks before and after Christmas were unusually rainy and cool in the Province of Guangdong; Shenzheng and Hong Kong were welcoming their guests for the holidays with an untypical palette of umbrellas and the unexpected boring need to wear top clothes. In the views of observers, the climatic "oddities" and "turnabouts" often precede or accompany the development of other natural or social anomalies on the way to their evolving into crises. But making such connections is only possible when one comes to think in the respective direction… And on the eve of 2003, in the hospitable Chinese South, the all-conquering tourist spirit was reigning supreme everywhere, despite the umbrellas, raincoats and jackets. One could hardly make his way in the midst of the thousands of visitors of the inimitable open-air Museum Park of the Nationalities; one could hardly make the full round of the exhibits in the Park of the World Cultural Monuments, or manage to see and assess the attractions in the Ocean Park and to squeeze in the motley crowd of people in the markets. Though atypical in terms of climate, on the eve of 2003 this picture had a single characteristic feature: overcrowding[3]. As it turned out much later, the

[3] I take the liberty here to say that the few hours' stay in the waiting room of Shenzheng railway station for one night in the period between Christmas and New Year in 2002, gave me the strongest ever self-awareness in my life of being an

lethal virus had already been raging in precisely this overcrowded place, precisely during these few weeks around Christmas[4]. And when the crisis strikes big agglomerations of people, its effect is more catastrophic and long lasting; the risk of catastrophes in overcrowded centers is a function of the degree of vulnerability – economic and social: of the incomes of the population, of the living standard, and of the administration[5].

Now it is already known that the mysterious disease was a kind of atypical pneumonia, whose infectious agent was a modification of a coronavirus of animal origin, not encountered so far among people (although there were speculations also about the role of the human mind in its creation). After comparatively long months of research it was assumed and later proven that the epidemic had started from mutated viruses in wild animals which had crossed over to people and had become modified. The assumption was that they spread through the inhaling tiny droplets secreted by the infected bearer during coughing or sneezing or via contact with secretions from a sick person. Its development, the starting point being the Chinese South – the Province of Guangdong – had been stimulated by the warm climate, the low sanitary standards and the gathering of a great multitude of people at one place, as well as by a set of local and general cultural features conducive to that development.

At the end of May 2003, the examination of secretions from wild animals, sold for food on the local markets in Guangdong, found that the coronavirus could be isolated from Civet cats. This suggested that the virus had crossed the biological barrier from cats to man, although it was claimed, even at the time, that they might not have been the only and the original source of the infection. Later on, the list of the animals carriers and original sources of the virus, extended to cover, besides the cats, some kinds of snakes, birds and mammals.

---

anonymous part of the million-strong crowd; no personal space for you or around you; people with children and luggage in their hands, on their backs, shoulders, heads, even on yokes, having stopped for a short or longer time midway. I fear that for some of them this had been the last stop.

[4] As a humble self-appeasement I later kept telling myself that the weather was, after all, favorable to man: no great heats, which could have made the way for the virus. And I did not feel like remembering the countless people in that waiting room, coughing, sneezing, spitting in the Chinese way all the time, eating, fighting their colds, snoring on the benches or noisily talking at a distance of just millimeters from my child, all the time trying – again in the Chinese manner – to touch it, push it a little or start talking to it.

[5] The Vulnerability of Cities: Natural Disasters and Social Science. 2003, p.212.

Obviously, the first essential risk factor conditioning the emergence of the crisis was the place these animals occupy in the everyday life of the Chinese from the South, their consistent presence in the culture. Another cultural characteristic on which ethnologists have to focus their attention in outlining the risk factors for the emergence of the syndrome are the eating habits of the Chinese, preserving to this day powerful traditional features difficult to change. Particularly important has turned out to be the stereotype practice of the Chinese "to meet" (instead of "to phone" or "to write") – the personal contact and direct intercourse as postulates in Chinese communication eased the powerful spread of the virus. Coming next are those behavioral cultural reactions, resting on the conception that if someone, a close relative or friend, has fallen ill, he should not be left alone. Complete spiritual comfort has to be offered to him through the presence, in close proximity around him, of all people near and dear to him. In this way a contingent of persons is created, potentially ready to be exposed to the virus, and who would allow in no way possible that their sick relative carry the official stigma in the form of the most stringent quarantine and isolation. And lastly, a series of strategic features of Chinese traditional culture in terms of psychology, which stood in the way of seeing the seriousness of the threat clearly and at an early stage: the silence kept around the earliest cases[6]; the long official silence of the government and the attempts to hide the problem until the first cases of the disease appeared in foreigners, and, among the Chinese physicians themselves, no voices were heard about making the problem public. Further on, I offer a few notes on some of these ethnological aspects of the SARS problem, precisely along the lines of the cultural niches, creating risk factors – a critical situation:

The southern part of China is famous for its large and overly rich markets of wild animals (though "wild" is not the exact term: these can be both game species hunted in the wild, and such bred with commercial purpose, but yet not "domesticated" in the anthropological meaning of the word, still preserving their biological characteristics from the stage of life while at large; a problem arises, however, from the fact that they are bred in direct proximity to people, practically living and being in contact with them without being "domestic"). This is one of the cultural peculiarities developed over the years and becoming an increasingly profitable tourist attraction, magnetically drawing rich Asian, Australian, American and

---

[6] In Shenzheng, according to foreign guests, people knew back in November 2002 that "a very bad flu is haunting the city" and wore masks, while the foreigners mocked them.

European tourists. From these reserves the wild animals are sent to restaurants and markets throughout the whole country: all kinds of wild dogs, polecats, badgers, and moles. As said, the Guangdong Civet cats, which acquired worldwide notoriety in connection with the laboratory examinations of the coronavirus, are assumed to have been the basic and initial carriers of the virus. (In the summer of 2003, about 10 000 of them were annihilated by a special decision of the Government).

The virus was directly transferred from animals to people through the food: the restaurants of the South are world famous for their dishes of game, each part of the animal having a peculiar significance, a way of processing and cooking, a place and time of serving on the table, a style of serving. The Guangdong dish known as "a dragon's fight with a tiger", for example, is widely popular and in great demand. It is cooked from the meat and bones of three kinds of poisonous snakes, Civet cats and various plucks of other such animals – stomach, liver, kidneys, and hearts. Superb gourmet delicacies, served in the so-called "exotic" restaurants are dishes of bear paw, of ant-eater (ant-bear), giant salamanders, etc. China has an extraordinary culinary subculture, indeed, which has been cultivated for more than 2000 years now. If offered as a main course, the pieces of meat are usually served raw or half-cooked. At home the meat pieces are also preferred raw, dipped in heated fat in a common bowl, everyone making the dish himself. The presence of meat of such animals in the everyday meals is not a luxury, unless, of course, it has been cooked with the renowned Chinese exquisiteness and scrupulousness. The prices of this kind of meat on the markets make it impossible for consumption by the common Chinese, but in this particular situation the very touch and getting close to it, without buying or consuming it, could be fatal, because very often, before being captured by the merchants, the animals had fed on the meat of dead animals. As a frequently repeated old Chinese joke runs, in the southern province of Guangdong, people "eat everything which has wings but airplanes; and everything which has legs but tables". By virtue of an old cultural habit, the shops selling animals for food are also the places, where the animals are slaughtered, should the customer wish to buy a whole animal or part of it. The sanitation is not of a very high standard, and remains and traces of animals' blood and bones can be seen. The same holds true for the backyards of "the exotic" restaurants, flowing from which are streams of blood of slaughtered animals which mix with the water of the ditches and small rivers flowing by.

When speaking about the culture of eating, we have inevitably to underline the fatal, in this case, yet obligatory model of collective meals. It is known that the Chinese take their meals out of common bowls, transferring the food with chopsticks into their own dishes. They, no doubt, like both their food and the way they consume it: sitting around a table with common bowls in its centre, each one pushing his chopsticks into the meal. The conception is that herein is a feeling of cordiality and good will, a feeling of togetherness, created by this way of taking meals, whereas the Western way of taking meals is judged by them to be "cold and clinical". Very often, particularly in the case of meat dishes, a bigger piece of meal is successively broken up in the common bowl by everyone in turn at the table; the remains and the scraps are left also on the table, next to the nearest one seated; people poke into the common bowls with their personal chopsticks, etc.

Quoted bellow are a few views, published by the mass media, of Chinese of different age and professional status on the questions of the culture of eating in the period of the crisis: Li Jian, Professor in sociology in Tsinhua: "It is very important for the Chinese to feel the friendly, joyful atmosphere of the common table; Besides, the Chinese traditional ways of taking meals strengthen the relations among people. But for that, quite a number of people would not even think of getting out for a meal". He also adds that the habits of eating, as an important part of Chinese culture, are not easy to change. For that matter when the SARS is surmounted and the fear driven away, quite naturally the people would reverse to their old practices. In his view, again, "the individual cutlery is part of the Western culture, emphasizing the individuality." Jen Uei, an owner of a restaurant in Beijing says: "I don't think that many Chinese like individual cutlery", and adds that some have an openly hostile attitude to it: "Many people come here with their families, but whoever serves individual cutlery at home? For many of the people, to offer individual cutlery means to be extravagant and impolite". Men Jan, a journalist, describes her going to a supper with friends soon after the epidemic had subsided, "My husband offered common cutlery and, judging by the faces of the people around, no one would have objected, even if they wanted very much to get individual cutlery". According to Djan Hun from the Chinese Hotel Association, the collective cutlery is appreciated as a symbol of the solidarity and friendship in the circle where it is practiced: "This is part of the Chinese culture"[7].

The stereotype having meals described above was the easiest to diagnose, point out and sanction as a carrier of the epidemic. Hundreds and thousands of restaurants, public catering establishments and common eating outlets were closed down; it was suggested to the people to buy

[7] Classic Catering Conflict: Culture vs. Caution. Jan 12, 2004.

foods from the shops and cook them at home. This dealt a blow, though not lethal, at the balance of personality-collective-culture. It turned out, however, that the closedown of the public catering establishments and the forced giving up of the traditional function of taking meals as a fundamental mechanism of communication, was accepted by society without any grumbling and resistance, without outstanding acts of disagreement and struggle, as was the case with other measures which violated the above-mentioned balance. The probable cause might be that the short-term change of the eating habits in the face of the epidemic in China had already had a precedent in 1989, when the country was struck by an epidemic of Hepatitis A. It had broken out in Hsiangt'an this time and, as this was often reminded by the mass media, for some time individual cutlery was enforced and practiced at the public catering establishments. But very soon, after the epidemic wore off, the traditional habits were resumed. That precedent probably made matters easier to bear in 2003, but the fact remains that, concerning the way of taking meals, the prevention measures were diligently followed without any protests at least until the end of the quarantine and emergency situation was declared.

The serious blow, spearheaded at breaking the millennia-old cultural stereotype and, therefore, received with powerful disagreement by society, came with the first wave of high mortality. The very first rumors about the unusual flu and the unexpectedly high initial mortality rate made the Chinese suspicious. However, during the months which preceded February 2003, they still turned willingly and voluntarily to physicians and did not object to the thorough laboratory analyses and examinations. The problem came to the surface when people with complaints started to disappear behind the hospital walls without any news whatever of them, without any information or allowed visits, telephone calls or any other form of contact. They just no longer appeared. That period coincided with the evening police and army "actions" undertaken by the order enforcement officials, dressed, as the hyperbolic saying went, "in space suits, and helmets", who led, pulled and dragged people with symptoms of the disease out of their homes and loaded them onto equally protected hospital cars. Horrible stories began to spread about the way of development of the disease and the fatal end. In these stories, the horror derived not so much from the state of the sick person, as from the loneliness of the individual before his end, completely unacceptable to Chinese culture. That is where the embryo of the risk is potentially hidden in traditional culture, which was to feed and stimulate the critical situation discussed.

In all cultures around the world there is a place kept for the closest relatives and friends that have passed away and ways of paying respect to the old people. In Chinese culture this attitude has developed to a degree assumed by some researchers to have become a hypertrophy. Far back, during the Middle Kingdom, the cult for the forefathers even strove to deepen, and death used to be celebrated as an apogee in such a way that everything in man's life on earth began to be interpreted as a prelude to the next life. In his book about the rites and rituals Confucius described in great detail a great number of ceremonies relevant to this cult and in this description can be seen the essential development and modification of the reverence for the forefathers. The road outlined by the emblematic Chinese philosopher in the development of the cult essentially shows the emergence of the family and clan pantheon and the funereal practices which are characteristic not only of Chinese culture. What is special here is the prominence of that cult and that it remained almost up to this day. Initially, the rites were performed exclusively with the aim of reviving the memories concerning the deceased relatives, of preserving their memory from oblivion. The expression "filial reverence", a stereotype for many cultures, also belongs to that time and to Confucianism, to be more precise. Throughout the centuries, numerous additions and further layers were accumulated onto that original idea, and, little by little, a belief developed that the founders of the clans, having gone to the world of spirits, were exerting from there influence on the lives and destinies of their descendants. Parallel to the gradual evolution of this belief, the consciousness of the importance of the cult for the close relations and friends who had passed away struck even deeper roots. Since the happiness and prosperity of the clan depended on its founder, the worship of the forefathers had to be performed impeccably and the relations linking the living and the dead should never be broken.

The simple memory of the deceased, what was probably the original form of the cult, has been replaced by worship performed diligently, now out of fear of those misfortunes and disasters, which may befall posterity, if it happened to neglect its deceased. In the Chinese world view, this idea unfolds in a detailed system of beliefs, starting from the assumption that man, in their view, has three souls. After death, one soul goes into "the dark world" where it gets into a situation close to the situation in which man had lived while alive; the difference is that the qualities of life in that world are lower than the earthly ones; life in the netherworld itself is deprived of the joys of life on earth. The second soul stays into the grave. And the third soul gets into a little piece of wood or a tablet, bearing the name of the deceased, which is reverentially kept by the deceased descendants. Sold to this day at the entry in the temples, open to visitors and pilgrims, are small red tablets (always two in number) onto which the names have to be written down of the people in whose memory, or for whose health, you want to pray. One of the tablets is placed at the foot of the altar, which is, as a rule, a huge figure of Buddha in some of the deity's variants, while the second tablet is taken home and is kept at a holy site in the house. Worship of the memory of the forefathers implies mostly worship of these latter two souls. This worship unfolds in a strictly observed rich set of rites and rituals: during the memorial holidays, the entire population leaves the settlement or the city and goes to pay a visit to the family tombs where, up to this day – and it even seems that nowadays the tradition is observed increasingly more strictly, – practices are performed detailed descriptions and analyses of which can be found in literature. In this way, according to the scholars, what was originally a common memorial service expressing filial love of descendants to their parents, over the centuries was transformed in China into idolatry, into a superstitious cult which "in many respects runs counter to common sense, because it creates odd, abnormal conditions of life, making the freedom of action of the living dependant on the disposition of their long deceased forefathers"[8].

The attitude to the aged and sick people, as a universal cultural reflection of the attitude to the deceased forefathers, is unambiguous and uniform in all strata of the Chinese people, among all educational, age, property, character and other categories of the Chinese population, in all regions and districts of the state: "The Chinese may bow or not to idols, they may believe in deities or show open atheism, and no one would pay any attention to this. But should just anyone happen to make light of the forefathers, and all the people, starting with his relatives and neighbors would rush at him with resentment. One of the strongest reproaches of the Chinese to those who have accepted Christianity is that in their new faith they neglect the forefathers".[9] It is customary for the adherents to Confucianism, as well as for the Buddhists and Taoists, to gather closely around the deathbed of their agonizing friend or relative and, on the arrival of death, to fill the premise with loud wails of grief and incantations; practiced to this day are also elements of the ancient custom of "recalling the soul" (passionate appeals to the soul of the deceased to come back to the living). Obligatory is also that each close friend or relative of the deceased should place in the hand of the deceased a lock of his/her hair as a token that after the reincarnation, they would be together again. Dying at home is considered a blessing in Chinese culture; even if someone happens to die in the

[8] D. Makgovan, I. Karostovets. Zhyzn i nravy starogo Kitaya [Life and Customs of Old China]. Smolensk, 2003, p.324.
[9] Ibid., p.75-76.

street, there is a special prayer and a requiescat prayer, and a requiescat service which ensure that the deceased's soul "settles" in the house alter, and remains there[10]. Up to this day the belief is firm and the requirement strict that in the fatal moment the person should be among his nearest and dearest, surrounded and cared after by them; and he/she should deliver his triple soul holding the hands of one's closest people, feeling their breath by his side; then he/she would give his blessing and light to illuminate their path, as well.

It was these beliefs and passions that the enforced official policy violated in 2003 in an effort to break them. Demanded was the prompt and categorical separation of the SARS patient from his environment and his close relatives and his closedown in an isolator for a period with unlimited duration – the policy of the stigma. This stigma enforced by the official authorities on the patients diagnosed to be carriers of the virus was unbreakable; it was enforced for months on end with all its severity and the intransigence of the law, the power and the weapons. At one point the Chinese realized that a person dear to them, who had just gone to the physician to have his temperature taken, might not come back at all, and his aura might fade away for them forever behind the walls of the quarantine wards. This became a nightmare for the average Chinese, because his cultural stereotype inherited and built up patiently and conscientiously, according to which the way a person leaves the earthly world was the most essential moment, was being destroyed. The reaction was immediate, sharp and unflinching – a reaction of resistance and refusal: the diseased began to be hidden, guarded equally strongly by close relatives and more distant people. As the months passed, this fatal non-compliance with a deeply rooted cultural model turned into a nightmare for the official authorities and they decreed forcible measures in order to overcome it, at least partially.

During that period numerous cultural stereotypes had to be violated and crushed by the prevention measures against the epidemic, and particularly by the stringency of the treatment technologies. But the most painful was the collapse of the funereal rites and rituals in Chinese culture. It is where the relationship of the person with the culture is especially distinctly felt: should it be lost, should the human dependence on the cultural ethos disappear, this inevitably produces a traumatic effect. The strength of the cultural memory and the ritual as a guardian, integrator and a system of security was subjected to threat, when the Chinese tried to undergo medical examinations and, for a number of reasons, was forced to distance him

[10] V.V.Maliavin. Kitaiskaya tsivilizatsiya [Chinese Civilization]. Moscow, 2000.

from the cultural habit and from those cultural values, which gave him enhanced social support. The observations on this situation suggest that culture ensures protection at a high cost: the strong attachment of the person to a certain way of life entails a deeper feeling of loss, when life is detached and set aside from the cultural stereotype[11].

Further, I would like to dwell on the inability of the common people in China, which showed during the crisis, to react and make decisions to their benefit – a fact, which was extensively commented in the analyses in the subsequent period as a cultural result, and, respectively, as a foreseeable complication. The problem lies in the way the Chinese bring up their children, the attitude to the collective and the responsibility they cultivate in them in their early years. What is deliberately developed in any person is the sense of absence of individual autonomy, absence of self-confidence in one's individual value and one's individual strength; and these are the very qualities which help people make decisions and realize them, confront the consequences and face the problems instead of avoiding them, or accusing the others, or resorting to demagoguery – in other words, these are the qualities which encourage a person to take risks in his/her endeavor to survive[12]. From a psychological point of view, people could have been capable of resolutely opposing every difficult process applying the mechanisms of individual autonomy and the strong feeling of self-sufficiency, because they would not have interpreted it as a personal threat to their identity, to their feeling of themselves. They would have, likewise, been more sensitive and accommodating, more respectful of the others.

It is expected that personal responsibility and determination to confront a critical situation would be stronger where the Western culture has exercised greater influence, and this, indeed, proved to be the case in the episode of SARS. According to the analyzers, the most prompt, decisive and responsible actions were taken by the governments and individuals in Singapore, followed by those in Hong Kong, Taiwan and, in the last place, in continental China. It was in China that the most eloquent examples of traditional Chinese personality models were displayed in action and precisely there, for that matter, the only adequate response to the epidemic were, understandably, provided by the political power. "These characteristics of the individual", one of the anthropologists observing the process wrote, "are not the result of contamination or something learnt from past

---

[11] M.W.DeVries. Trauma in Cultural Perspective. - In: Van der Kolk, B.A., McFarlane, A.C., Weisaeth, L. (eds.). Traumatic Stress, New York, 1996.

[12] J.Polakova. The Struggle for Human Dignity in Extreme Situations. - In: The Vulnerability of Cities: Natural Disasters and Social Science, 2003.

experience, nor are they some ethno-psychological characteristic of the Chinese of the Han group. They are created in an environment of well-wishing, in which each person grows up and develops in the family and at school".

Indeed, the basic feature of character encouraged and cultivated in children in China even today is the strong group consciousness or a sense of group identity, rather than clear, defined and strong autonomy and independence. And this is by no means the decision of a certain family, but comes as a result of traditional practices of upbringing which act in two directions: no one hand, it is the extraordinary pampering of the kids with excessive love and, on the other hand , a very close and rather strict control. This is a form of upbringing which has as a result highly harmonized social interdependences between the individual and the group. That is why, when the harmony within the group is lost because of a crisis, the individual is shocked, horrified and confused, there being nothing else he can rely on. So, on the Chinese scale of personal values, the strong individual's feature of making decisions and taking firm actions for their implementation (features, which may infringe on the harmony), has been substituted for by hiding, overlapping, accusing, delaying and lack of determination. Some of the reasons why the crisis of the disease was not resolved promptly and categorically can also be sought in these features of the individuals encouraged through their education in childhood.

Another effect of the Chinese upbringing traditions observed to this day, aiming at the formation of a high degree of group identification and absence of personal consciousness, is the individual's inability to see the person by one's side as an independent subject. For that matter, the compassion and the feeling of social support, as well as relying on the higher good will and solution, were staked for the needs and security of the individual personality. This is the reason, once again, why in China the only possible winning step was the adequate work of the institutions of power[13]. It is known that they had delayed their reaction and had made an attempt at covering up the truth until the moment when its bringing to the fore was already inevitable. The anthropological observations here confirm the idea that social adjustments during a crisis "are not radical, abnormal breaks with customary behavior; rather they extend ordinary conventions..."[14] In accordance with this specific feature, when the scale of the problem came

---

[13] T.Palakudiyil and M.Todd. Facing up the Storm – How Local Communities Can Cope with Disaster: Lessons from Orissia and Gujarat, 2003, p.122.

[14] W.Torry. Morality and Harm: Hindu Peasant Adjustments to Famines. *Social Science Information*, Vol.25, 1986, p.126.

to be discussed out loud, the collectivistic spirit and mass discipline, the fear of punishment for the group and the subordination of its prestige began to gradually stimulate the healing process until its completion.

When analyzing the discussed situation from the distance of some time passed, I had before my eyes also the experience of a reaction to similar events in my own country. A brief presentation of these events might contribute to a better understanding of the anthropological cohesion risk-crisis-culture in the data presented. A slightly different situation, though caused by similar sociological and psychological concerns as the SARS, was faced by the small Balkan country of Bulgaria when threatened by a Bird Flu later in the new century. Hit by this threat in 2006, Bulgaria was forced to react accordingly. What was the official and unofficial reply there, and how it could help interpret anthropologically the much bigger and dangerous spread of SARS in China a few years earlier?

As it has already been mentioned, both situations – in China 2003, and in Bulgaria 2006, differed. In the first place, in 2006 the Bird Flu was anything but news. Actually, the knowledge of the virus was a few decades old, and its dark reputation as the alleged killer of 50 or even 100 million people in 1918 was well remembered[15]. The ghost of the disease haunted the open public discourse in Europe in the year of SARS outbreak in Asia (2003), when the article of the famous European virologist Prof. R. Webster was published in *American Scientist* entitled "The world is teetering on the edge of a pandemic that could kill a large fraction of the human population"[16]. He referred there to "influenza caused by viruses in birds", commonly called Bird Flu, and saw it as a major world threat to possibly as many as billions of people. It was also known that all Bird Flu viruses belong to the A-Flu virus from the Orthomyxoviridae-family, and at the time it was also well-known that Avian Influenza or Avian Flu (sometimes called H5N1) infected birds, including chickens, poultry and wild birds such as ducks, swans, etc. The article was making the point that, although most Bird Flu viruses can only infect birds, these viruses might change, or mutate to forms that more easily infect humans. As a result, this could lead to a pandemic, or a worldwide outbreak of the illness. This article of Prof. Webster in 2003 aroused public consciousness about the possibility that H5N1 may also mutate into a strain capable of human-to-human transmission.

---

[15] G.Kolata, Flu: The Story of the Great Influenza Pandemic of 1918 and the Search for the Virus that Caused It, New York, 1999.

[16] R.Webster. The world is teetering on the edge of a pandemic that could kill a large fraction of the human population. *American Scientist*, issue March-April, 2003.

This was the second difference: records and scientific knowledge about the virus were present. This picture is more typical than that of an epidemic which is totally unknown and unexpected factor. The epidemic has occurred in the past and human memory or local traditions can recall it. The history of the Bird Flu spreading to humans was itself well documented: The first case was registered in Hong Kong in 1997 during a Bird Flu epidemic; seriously ill were 18 people, 6 of whom died. At the end of 2003 and the beginning of 2004 outbreaks of the epidemic were registered among the birds in 8 Asian countries: Cambodia, China, Indonesia, Japan, South Korea, Thailand and Vietnam. The consequence was the death of 100 million birds there, killed both by the disease and in the human attempts to prevent it. So, provided with the said historical background, neither the farmers, nor the local administration in the areas concerned remained passive actors in the face of possible or actual pandemic. In March 2004 it was claimed that the epidemic was under control. But in June 2004 new cases of Bird Flu appeared again in some Asian countries – Cambodia, China, Indonesia, Malaysia, Thailand and Vietnam.

These two specifics eliminated the risk of a sudden outbreak, minimizing in this way the "surprise-effect" as an eventual excuse of delayed reactions. Besides the lack of incompetence, scientific awareness and good analytical knowledge of the disease and its consequences, one important fact was decisive – and this is the third distinctive feature of the situation there: Bulgaria was one step away from its accepting in the EU. Everything its officials and the whole society were doing was carefully watched and evaluated in Europe, which meant that a wrong step in that situation of a world-wide concern could have been dearly paid back; hence, the extremely responsible and meticulous attention to the Bird Flu cases.

Actually, the attention – both public and official – had been directed to the spread of Bird Flu in the countries around Bulgaria since the middle of 2003: in the second half of the year, all dead birds from poultry farms situated in the area of the Danube and Black Sea were inspected – 500 blood analyses were done, and 154 domestic and over 100 dead wild birds were tested negatively for Bird Flu. As for example, in December 2004, sick birds were found in Makala Yadere, village Ridina, the province of Djebel, in Southeast Bulgaria. The presence of Newcastle disease was confirmed, but even if the doubts of Bird Flu were not proved, all the birds were exterminated. By that time, all import of poultry, eggs, etc., from the countries with Bird Flu cases had already been forbidden; foreign experts were allowed to inspect certain suspicious areas and animals, both wild and domestic[17]. After October 8th, 2005 Bulgaria felt directly threatened because cases of sick birds were registered in Romania and Turkey, and in Turkey there were also sick people diagnosed with Bird Fly. By that time, 3 dead birds had been found in the area of Pleven (Central North Bulgaria).

The early weeks of the new 2006 were the most anxious time for the Bulgarian authorities and citizens[18]. Most affected were, indeed, the territories in the immediate neighborhood of Romania and Turkey, and opinions like, "Now being sandwiched between Romania and Turkey, yes I am worried about the spread of this virus," were often heard in the on-line debates – fears that the outbreak of the infection could cross the borders into Bulgaria caused panic especially in the households that raised domestic birds. All veterinary departments and units of the Civil Defense were supplied with Tamiflu so that the possible spread of the disease could be prevented. A close eye was kept on the overall situation: each new case in Europe was announced and discussed in Bulgarian media.

In the year 2006, Bulgaria actually had 2 instances of fear and, following it, mobilization regarding the Bird Flu threat: First, at the beginning of February, a dead wild swan was found on the territory bordering with Romania, in Vidin, Northwest Bulgaria (03.02), and a week later two other birds were found in Northeast Bulgaria, respectively in Dobrich and in Varna (09.02). From the very beginning, it was presumed that these were the first Bulgarian cases of Bird Flu, as it was announced by the agricultural minister N.Kabil. Until these cases, all dead birds found and tested, were said to have died from the so-called "Newcastle disease" (or pseudopest), or from cold and hunger.

A bit earlier, on January 12th, a National Pandemic Readiness Plan had been adopted by the Government, worth 3.7 million Leva (1.8 million US dollars). This plan included also farmers' compensations for exterminated birds, and this was the proper sanction to ensure the farmers' goodwill-attitude. Starting in February, the first weeks' panic was followed by a considerably quick and more adequate reaction by the authorities: detailed Measures for Prevention were announced by the Crisis Headquarters. It clearly stated that "the contamination comes from birds – domestic (mostly hens and turkeys) and also wild (mostly water birds – turkeys and ducks); the turkeys could be contagious even if completely healthy…" But in almost all cases, those who were infected with H5N1 had extensive physical contact with infected birds. Thus the category of the potential

---

17 www.medicalnewstoday.com/articles/37663.php.

18 www.bird-flu-center.com/category/europe/bulgaria.

victims was still limited to people dealing with birds, despite the threatening warning from the past years, the statistics being, indeed, a reason for anxiety and fear: about 60% of humans known to have been infected with the H5N1 had died from it – a number which was repeatedly announced.

When this information was loudly discussed, and the high lethality of H5N1 and its significant ongoing mutations were supposedly understood by the people concerned, the situation changed with a higher social discipline and voluntary obedience shown to the orders. Meanwhile, billions of dollars were being spent worldwide on research on H5N1 and preparation for a potential influenza pandemic: during the crisis in 2003-2006 at least 12 companies and 17 governments were developing pre-pandemic influenza vaccines in 28 different clinical trials that could avert a deadly pandemic infection, if the trials were successful[19]. One additional problem was that a full-scale production of a vaccine that could prevent any illness at all from the strain would require at least three months after the virus erupts, which meant that fatalities could not be completely avoided.

The second instance of serious Bird Flu-threat was in the summer of 2006, after some dead domestic birds were found in Southeast Bulgaria, in Djebel (20.07). It was confirmed that the tests did not show the presence of the highly pathogenic virus of Bird Flu, but still the measures were drastic. The official statistics show that 5 months after the registration of the first positive case of Bird Flu, over 7700 examinations of dead birds were made, and only 4 of them were positively diagnosed with H5N1[20]. The enhanced attention and monitoring of the birds' migration and the behavior of the migratory birds, as well as the prevention of the contacts between the wild and domestic birds which could transmit the virus, are the main reasons for the absence of cases of Bird Flu in Bulgaria. The requirements for extermination of domestic birds were felt, though, unpopular and unwelcome.

At this point, we could draw a possible anthropological conclusion, which should be taken into account if other similar situations appear: Bulgarians in almost all regions of the country raise poultry as their traditional livelihood. Deeply rooted in their traditional value system is a sincere concern about their live stock, especially birds; they have always been emotionally attached to the health, good looks and the qualities of the domestic birds; birds – that is, hens, roosters and chickens, ducks, geese, turkeys, etc. – have always had their place and role in the calendar and

[19] www.einnews.com/bulgaria/newsfeed-bulgaria-bird-flu.

[20] www.bird-flu-center.com/category/europe/bulgaria.

family customs, and their presence in the rituals has significant symbolic meaning usually associated with prosperity, health and good harvest[21]. Accordingly, farmers used to successfully treat the common and even more severe diseases in their poultry farms: as elsewhere, if the disease is of a limited scale and duration and happens relatively frequently, simple adaptive measures have been generated and adaptive practices developed through which risk and damage are minimized. Therefore, forcing the farmers to exterminate their birds just because of a shadow of a strange disease never personally experienced by the farmers was an unpopular measure, and many farmers tried to avoid it. What is more, on the eve of Bulgaria's acceptance into the EU, poultry was thought to be of serious export value for them; during the previous years the farmers had spend a lot of financial resources for improving the conditions of domestic stock's breeding, raising and producing the respective products. And by March 2006 they had already begun loosing money, the price of the broilers having gone down to the record of 2 Leva/kg, so the State had to ensure 3 million BGN Leva as compensation because of the low prices. Therefore, the farmers just couldn't agree over night to kill their birds and hopes for an expected better wellbeing very soon. And they began refusing the extermination of the birds and searching for different ways to disregard the announced rules.

In addition, the negative reaction to the measures announced, although not frequent, was grounded on the statement of the Direction of the Crises' Headquarters at the beginning of January that, "It is believed, that a certain risk exists when poultry products are consumed which have not processed through heat..." Raw mead is not favorite in the Bulgarian kitchen at all; neither in the traditional, nor in the modern kitchen and diet can one find a recipe or a dish containing raw meat of any kind. Therefore, the farmers were taking this as an unspoken excuse not to obey the requirements for killing their stock. It was even stated in the Directions that "there are no data of people getting sick after eating thermally processed poultry – meat or eggs, because the Bird Flu viruses die having undergone such procedures".

The officials fought back: starting from January 23th, within less than 2 months, 87 farmers had to pay penalties for not obeying the directions for extermination or not following the safety requirements. The possible conclusion is that either the compensations were not good enough for the farmers or they lacked proper information about the seriousness of the

[21] Etnografia na Bulgaria [Ethnography of Bulgaria]. Vol. 2, Material Culture. Sofia, 1983, p.28-32.

threat. Whatever the explanation, the disobedience was not dramatic and sharp (it was expressed on a domestic households' level only), and does not in the least bit resemble what was observed in China and Asia in 2003.

Examining the government and the community reaction, two characteristics are obvious: it was 1) transparent, and 2) mobilizing. From the very beginning, and especially from February on, a full transparency of the measures taken was not only present but also widely announced and demonstrated. The central government's steps, as well as those of the local officials in the regions concerned, were presented for evaluation to the people inside, and to all foreign experts and officials. In comparison to what I observed in China a few years ago, the attitude of the people in Bulgaria was closer to a dynamic organization of prevention than to a panic. Some comments in the on-line forums and some emotional discussions about the possibility of the birds coming from Turkey transferring the Flu were the sign of a fear, but it was not dominating. However, it was not the deadly threat that forced this reaction. More likely, it was the approaching of the New Year's Day, the expected first day of the Bulgarian membership in the EU. It is also proven by the fact that each single measure taken was loudly and widely announced in all the media and discussed. Each day the newspapers and the radio- and TV-broadcasts began their emissions with the news about what was done in regard to Bird Fly prevention. The on-line news also gave a strict follow-up, even minute-to-minute info about this. The Bird Fu threat was eliminated thanks to the measures taken at the right time – sometimes after, but also in advance – the careful watch of the neighboring situation and the following of all requirements by the EU – literary and fully.

The two presented cases are not compatible even in their major parameters. However, the descriptions and thoughts shared, though sketchy and concise, give grounds to formulate a few conclusions regarding the relationship risk-crisis-culture from anthropological point of view. In the first place, it is clear that in the streamlined chain of traditional culture and the values suggested by it (not only Chinese and Bulgarian but any other), precisely because of its systematic character, there are "weaker" units or such which in the course of cultural creation draw from nature and practice and thus become vulnerable to the strikes of crises. They create a vacuum, deprived of cultural order (a space of real or supposed chaos), which carries the signs of risk and is ready to fit in and give grounds to the symptoms of the crisis. Three such units have been discussed in this paper, drawing mostly on Chinese data as examples in one specific culture: specificity of the diet and the way of taking meals; some specific features of the beliefs and practices associated with death and the funereal rites and rituals, as well as the interpretations of the trend of traditional upbringing in the Chinese family. Today, some time after the strike and its echo have faded out, it seems hard to explain why scholars and politicians had not thought of focusing on and acting according to these specificities. We, the ethnologists, are obliged to show them these specific features in order to facilitate society and the individuals in their struggle to survive, which is by no means an easy task.

The data shared contribute also to a discussion on whether some illnesses exist only among members of a particular culture. The idea of a positive answer was given by investigations precisely on Asian cases[22]. As it has been proven in the present study, even as far as such highly specific diseases as SARS and Bird Flu are concerned, certain complex illness conditions have to be present so that the disease can develop. These conditions are certainly "culture-bound", but their social and psychological prerequisites do not differ significantly from any other illness conditions. They are, therefore, "not so much peculiar diseases but distinctive local cultural expressions of much more common illness conditions that can be found in any culture"[23]. The cases discussed confirm the idea that there are indeed cultural conditions of pre-disaster vulnerability that predispose some communities to experience disaster more often than others, or to suffer a higher disaster impact than others, and they should be studied by the anthropologists. However, these conditions are not decisive in our days. Therefore, full-scale social scientists' joint efforts are necessary for a responsible disasters' analysis.

So, even if the ability of local institutions to mitigate the impact of a disease forms the specific profile of each of the pandemic-crises situations, the most important conclusion concerns the way in which this concrete situation is evaluated in the official world's view. All public perceptions about risk and disasters are shared constructs, not depending on a particular ethnicity or culture; therefore, understanding how people think about and choose between risks must be based on the study of culturally-informed values as well as their social and political context, which is the anthropologists' task. The sociopolitical context of the reaction to a real

[22] Observations on a disease called *phii pob* in some northern Tai villages, was viewed by the Thai Physician Sangun Suwanlert as restricted to only this area, and it was diagnosed as a "culture-bound syndrome" that can only occur among people who share rural Thai cultural values and beliefs: Sangun Suwanlert. Phii Pob: Spirit Possession in Rural Thailand. – In: W. Lebra (ed). Culture-Bound Syndromes, Ethnopsychiatry, and Alternate Therapies, *Mental Health Research in Asia and the Pacifi,* Vol.4, 1976.

[23] R.Hahn. Sickness and Healing: An Anthropological Perspective, 1995.

disease (in the case of China's SARS) or the "imagined" or possible and shadowed disease (in the case of Bulgaria's Bird Flu) was the frame and the force that molded and directed the reactions. In both cases the concerns, close attention and categorical demands of the international community and its official institutions were the motor for successful mobilizational and organizational steps. Since, as it has been shown above, the epidemics challenge the structure and organization of society, the socio-political outside force minimized, to a certain extent, the possible effects of some cultural stereotypes that could open the niches for development of the diseases. Nevertheless, these stereotypes should be studied in order to react intelligently and properly according to a wider cultural and sociological context, when similar situations arise elsewhere in our world. The cases discussed contribute to the idea that in our present-day global society, where the overall state reputation builds on its official political accordance with the collectively and publicly taken measures, the close international monitoring and frequent interference of authorized foreign expert institutions can help shape more effective responses to natural, biological, technological and social catastrophes. As a methodological assumption, the study shows that anthropologists should approach pandemics within their social, cultural, political, economic, and environmental relationships, with a special attention paid to culturally created disasters' niches, adaptations and responses – everything that can help prevent the disaster and lessen its impact.

References

Classic Catering Conflict: Culture vs. Caution (Jan 12, 2004).

DeVries, M.W. Trauma in Cultural Perspective. - In: Van der Kolk, B.A., McFarlane, A.C., Weisaeth, L. (eds.). Traumatic Stress. New York, NY Guilford Press, 1996.

Etnografia na Bulgaria [Ethnography of Bulgaria]. Vol. 2, Material Culture. Sofia, 1983.

Hahn, R. A. Sickness and Healing: An Anthropological Perspective, 1995.

Kolata, G. Flu: The Story of the Great Influenza Pandemic of 1918 and the Search for the Virus that Caused It. New York, Touchstone Books, 1999.

Kolev, N. Kritichni situatsii i narodniya mirogled u balgarite [Critical Situations and the Popular World View of the Bulgarians]. Veliko Tarnovo, Faber, 2003.

Makgovan, D., I. Karostovets. Zhyzn i nravy starogo Kitaya [Life and Customs of Old China], Smolensk, 2003.

Maliavin, V.V, Kitaiskaya tsivilizatsiya [Chinese Civilization], Moscow, 2000.

Palakudiyilp T. and M.Todd. Facing up the Storm – How Local Communities Can Cope with Disaster: Lessons from Orissia and Gujarat, 2003.

Polakova J. The Struggle for Human Dignity in Extreme Situations - In: The Vulnerability of Cities: Natural Disasters and Social Science. Mark Pelling, 2003.

Sangun, Suwanlert. Phii Pob: Spirit Possession in Rural Thailand. - In: Lebra W. (ed.). Culture-Bound Syndromes, Ethnopsychiatry, and Alternate Therapies. *Mental Health Research in Asia and the Pacific*, Vol.4, The University of Hawaii Press, 1976.

The Vulnerability of Cities: Natural Disasters and Social Science. 2003.

Torry, W. Morality and Harm: Hindu Peasant Adjustments to Famines. *Social Science Information,* Vol. 25, 1986, p. 125-160.

Webster, R. The world is teetering on the edge of a pandemic that could kill a large fraction of the human population. *American Scientist*, Issue March-April, 2003.

Internet Sources

news.avianinfluenza.info/bird_flu_bulgaria_news.html

www.einnews.com/bulgaria/newsfeed-bulgaria-bird-flu

www.medicalnewstoday.com/articles/37663.php

www.bird-flu-center.com/category/europe/bulgaria

Abstract

This article was inspired by the idea of collecting and studying some of the numerous examples of crises, disasters and extraordinary situations in human society, to theorize on them and to build possible models of the emergence, operation, prevention, reaction to and consequences thereof from anthropological point of view. In these pages thoughts have been shared about such social crisis which broke out in 2002-2003 in Southeast Asia: the spread and combat against the syndrome of atypical pneumonia. An attempt has been made to scrutinize those "niches" in Chinese culture which allowed the evolvement of risk factors, while they, in turn, caused the emergence and deployment of the epidemic crisis. To interpret the SARS situation more adequately, data from another supposed pandemic-case were involved – i.e., Bulgaria's Bird Flu threat in 2006. The author's conclusions reveal the cohesion between socio-cultural and socio-political factors in these two situations, and their shifting roles in allowing the crisis to happen, and in hindering/or helping it to be solved.

Elya Tzaneva, Assoc.Prof. and scholarly Secretary of the Ethnographic Institute and Museum, Bulgarian Academy of Sciences; PhD in Ethnology (Moscow University); PhD in Sociology and Social Anthropology (University of New South Wales, Sydney. Research interests: in the field of the Theory of Ethnicity, Nation and Identity; Traditional Family Customs and Fictive Kinship. Author of studies and articles in the field, as well as of the monographic work: "*Interpreting Ethnicity*" (2000); also "*Bulgarian Ethnography*" (2000, 2006, 2008).

ABBREVIATIONS

AEIM	Archive of the Ethnographic Institute and Museum, Sofia
BAS	Bulgarian Academy of Sciences, Sofia
BE	Bulgarian Ethnology. Ethnographic Institute and Museum, Sofia
CASS	Chinese Academy of Social Sciences, Beijing
CNY	Yuan (Chinese currency)
MUC	Munzu University of China (former CUN – Chinese University for Nationalities), Beijing
EB	Ethnologia Bulgarica, Sofia
EI	Ethnographic Study of the Village of – In: RKS [A Manuscript of Students' Course Work by the Dept. of Philology, Sofia University].
EIM	Ethnographic Institute and Museum, Bulgarian Academy of Sciences, Sofia
IEIM	Studies by the Ethnographic Institute and Museum. Sofia
IMIR	Center for Minority Studies and Intercultural Relations, Sofia
INEM	Studies by the National Ethnographic Muzeum. Sofia
Lv	Leva (Bulgarian currency)
mu	Traditional Chinese unit of area = 0.067 hectare = 0.164 acre
SbNU	A Collection of Traditional Culture and Folklore. Sofia

CONTRIBUTORS' SCHOLARLY INSTITUTIONS AND CONTACTS

Institute of Ethnology and Anthropology, Chinese Academy of Social Sciences
Bldg 6, 27 Zhongguancun South Street
Beijing 100081
China
Tel: 8610-6893-2337

Hao Shiyuan - hsyuan@263.net
Zha Luo - zhaluo@sohu.com
Liang Jingzhi - liangjz999@yahoo.com.cn
Fang Sumei - cbfls@yahoo.com.cn

School of Ethnology and Sociology, Minzu University of China (former Central University for Nationalities)
27 Zhongguancun Nandajie
Beijing 100081
China
Tel: 8610-6893-2324; Fax: 8610-6846-6568

Yang Shengmin - yangshengmin@sohu.com
Mingxin Liu - liu6949@126.com
Hou Yuangao

Ethnographic Institute and Museum, Bulgarian Academy of Sciences
1000 Sofia, 6a Moskovska Str
Bulgaria
Tel: 003592-8052613; Fax: 003592-9884209

Rachko Popov - eim_ban@abv.bg
Radost Ivanova – radost_ivanova40@abv.bg

Iva Kyurkchieva - iva filipova (ivaphilipova@yahoo.com)
Elya Tzaneva – etzaneva@gmail.com

Sofia University St.Kliment Ohridski
Faculty of History, Dept. of Ethnology
1504 Sofia, 15 Tsar Osvoboditel Blvd., room 26A
Bulgaria
Tel: 003592-9308-356

Tanya Boneva - tanja_boneva@hotmail.com

St Cyril and St Methodius University of Veliko Tarnovo
Faculty of History, Dept. of Ethnology
5003 Veliko Tarnovo, 2 T.Tarnovski Str, room 605,
Bulgaria
Tel: 00359-62618-317

Nikolay Kolev - mbox@uni-vt.bg